Green Energy and Technology

Alan J. Sangster

Energy for a Warming World

A Plan to Hasten the Demise of Fossil Fuels

Alan J. Sangster, PhD, CEng, FIET
Heriot-Watt University
School of Engineering and Physical Science
Edinburgh EH14 4AS
United Kingdom
a.j.sangster@hw.ac.uk

ISSN 1865-3529 e-ISSN 1865-3537
ISBN 978-1-4471-2535-8 e-ISBN 978-1-84882-834-6
DOI 10.1007/978-1-84882-834-6
Springer Dordrecht Heidelberg London New York

British Library Cataloguing in Publication Data
A catalogue record for this book is available from the British Library

Softcover reprint of the hardcover 1st edition 2010

Cover design: WMXDesign, Heidelberg, Germany

Printed on acid-free paper

Springer is part of Springer Science+Business Media (www.springer.com)

To Emily

I trust you will not have cause, one day,
to castigate my generation for leaving
an impoverished planet for yours

Preface

In December 2007 I was motivated, by something I had read relating to the environment, to submit to the editor of a long established Scottish newspaper, namely the Herald, a letter making some comments on global warming, which at the time I felt needed to be expressed. It contained the following paragraph:

> It troubles me that the news media, politicians, industrialists, economists and even some scientists continue to 'green-wash' the situation by propagating the lie that renewable sources of power will allow 6.5 billion people, growing rapidly to 10 billion, to pursue Western style energy wasteful modes of living, while at the same time protecting the planet. I suspect that even if every suitable pocket of land on the surface of the planet were covered with windmills, solar panels and bio-fuel crops, and if every suitable sea shelf, estuary and strait were furnished with windmills, wave machines and barrage systems, we would still have insufficient power from renewables to accomplish this.

Since submitting it, I have been exercised by niggling doubts as to the extent to which this statement is fully supported by the scientific and engineering evidence. My 'gut feeling' – not an instinct I like to rely on too much as an engineer – on the basis of my long acquaintance with electrical systems, and of wide reading on the subject of global warming, is that it probably expresses a grain of truth about the exaggerated claims for 'renewables', not by those 'at the sharp end' developing these renewable systems, I hasten to add, but by those with a vested interested in unimpeded economic growth. The 'spin', which largely amounts to unsubstantiated assertions made repeatedly in certain organs of the 'media', in effect suggests that renewable resources can provide a complete replacement for fossil fuels, when they eventually run out, or preferably, are locked below ground before they do. If we assume that a post fossil fuel era will arrive, sooner or later, the implication is that for the foreseeable future energy supply will not be constrained, and hence that 'business as usual', particularly in the industrialised world, is possible. I should note that here, and throughout the book, the term renewable energy implies energy diverted to human use, which is endlessly available as a result of daily solar radiation passing through the atmosphere and striking the surface of the planet. This diversion does not add to or subtract from the Earth's energy balance and is thus sustainable.

Some well established global warming arguments, which suggest that business-as-usual is not an option for mankind, are revisited in Chap. 1. The exhortations emanating from these arguments urging the global community to drastically cut fossil fuel usage, and to expand energy supply from the so called 'renewables', are also reconsidered from an electrical engineer's perspective. However, despite the expressed fears, the commonplace presumption appears to have developed, for whatever reason, that the amount of power that mankind can potentially harness from hydro, wind, wave, sun and other renewable resources, is more than large enough to assuage future demand levels. While the levels of potential global power consumption, which are well documented, usually in official reports, are generally accepted as being reliable, the presumption of unlimited power from renewables is like saying that since we have enough land to grow all the wheat we need, the future global consumption of bread will be satisfied. Just because enough land may be available it does not necessarily mean that it will be allocated to the growing of wheat, or that enough wheat will be grown, or that grain will be available where it is needed, or that enough bread will be baked where it is most required. Such a statement of potential capacity doesn't really get you very far. Competing interests will inevitably interfere. What we need to compare is electrical power that can reasonably be delivered to consumer sockets (after taking account of land suitability, land use, losses in the electrical generation and transmission systems), with the rate at which fossil-fuels are being consumed worldwide, to get a more realistic appreciation of the extent to which renewable capacity and global demand are likely to converge.

Here the issue has been examined from a more firmly focused engineering perspective than appears to have been attempted elsewhere. By taking a closer look at the original, readily available, undoctored power and energy data for renewable resources, it has been possible to construct, a coherent and comprehensive, scientific account of the current situation, vis-à-vis the potential capacity of alternative power supplies. From this firm knowledge base, an attempt has then been made to develop reliable engineering predictions of the exploitation potential of each of these sustainable resources in a 30–40 year time frame. In so doing it has been necessary to assume that we can depend on technology that is either currently available or is presently under development, and is therefore capable of being brought on-stream in this timescale. Also, by relying on well established electrical engineering laws, techniques and data, the computational process has, hopefully, allowed us to arrive at firm estimates for the power, which might realistically be transmitted to global consumers from these sources.

As far as has been possible I have conducted the energy assessment exercise with my 'engineering hat' firmly on, and hopefully much of the content reflects this. However, any book impinging on global warming, the truth or otherwise of anthropogenic forcing, and the problems of weaning mankind off its dependency on fossil fuels, is inevitably dealing with intensely economic and political issues. Consequently, it has obviously been difficult not to enter this political debate to some extent, no matter how tangential some of these issues may be to the main thrust of the book. Where I have done so, intentionally or unintentionally, I can only hope that the contributions are justifiable and helpful. The approach will probably be dismissed, in some quarters, as being economically naïve, but given the events of 2008 which suggest that 'economic science' is on the point of unravelling, who knows what now constitutes sound economics? Notwithstanding the intentionally narrow scope of the exercise, the engineering logic has led inexorably to a global perspective on renewable power supply and transmission, which has some surprising and uncomfortable ramifications for mankind. While several contributors to the debate have hinted at some of these consequences, I am not aware of any alternative assessments of the issues of global electrical supply and demand in the post fossil fuel era, which also highlight the potentially awkward implications that are lying in wait for advanced societies in making the transition to renewables.

Within the main chapters of the book I have attempted to furnish enough in the way of electrical engineering fundamentals to provide a primer for the reader to help him/her to appreciate the following: how renewable sources of energy can be exploited to provide electricity: how the electricity is generated and transmitted: what the constraints are: where the limits to the exploitation of renewable resources lie: how we can overcome intermittency of supply. While we shall need some basic physics and some elementary electrical engineering concepts to intelligently develop our arguments, this is certainly not an electrical engineering book in the college text sense. It contains no electrical engineering science beyond a very basic, school science, level. A good understanding of energy and power relationships, which are often poorly understood by non-scientists, is key to being able to assess or question the claims of the energy industry, particularly in relation to 'renewables', and to reach as wide an audience as possible the book attempts, largely through analogy, to illuminate these relationships in Chap. 2. Nonetheless, engineering and scientific concepts are most precisely expressed through mathematics, and for those who did not turn their backs on the subject at an early age, some relevant equations are provided in the referenced 'notes'. Renewable sources of power and their exploitability are evaluated in Chap. 3, while the enabling topic of massive energy storage is dealt with in Chap. 4. The final chapter is Chap. 5, in which some engineering based conclusions, and I stress 'engineering based', tinged with some unavoidable, but hopefully helpful, personal observations, are presented, with the aim of examining the manner in which the technological transition might possibly proceed, to a world in which electricity is supplied entirely from renewable resources, as they become the only source of power that mankind can safely access.

Naturally all views, assertions, claims, calculations and items of factual information contained in this book have been selected or generated by myself, and any errors therein are my responsibility. However, the book would not have seen the light of day without numerous personal interactions (too many to identify), with family, with friends, and with colleagues at the Heriot-Watt University, on the topic of global warming. So if I have talked to you on this topic, I thank you for your contribution, and the stimulus it may have provided for the creation of this book. I would, also, particularly like to thank my son Iain (Sangster Design) for one of the illustrations, and the members of staff at the Heriot-Watt University library, who have been very helpful in ensuring that I was able to access a wide range of written material, the contents of some of which have been germane to the realisation of this project.

Edinburgh, Scotland 2009 *Alan J. Sangster*

Contents

Chapter 1
The Context and Corollaries

A billion could live off the Earth; 6 billion living as we do is far too many, and you run out of planet in no time.

James Lovelock

Fixing the problem (of global warming) will not cost us the Earth, whereas not fixing it will certainly cost us the Earth.

John Ashton

1.1 Weather Warnings

Are human beings and human activities having a negative influence on the ecology of the planet? The population of the globe is now (in 2008) at 6.7 billion, and with a sizeable proportion of these billions living energy-profligate lifestyles it seems increasingly difficult to deny the fact – although many still do! If you have ever viewed night time satellite images of the Earth, when the surface is not shrouded in cloud, the evidence of the presence of mankind is staggering. Excess light now splashes over virtually all of the industrialised nations of the globe. Cities, towns, villages, motorways, trunk roads and other travel routes are easily identified. If carbon emissions, and carbon dioxide molecules, could be 'seen' by human eyes in the way we detect photons would we discern a similar picture? Roughly 80% of the world energy comes from burning coal, oil and gas. Immense benefits have clearly accrued to a growing section of mankind from the combustion of these fuels, which are derived from the fossilised remains of plants and animals, as a result of being compressed below ground for hundreds of millions of years. A veritable treasure trove! But re-releasing all this buried carbon into the atmosphere is not without cost. Evidence is growing that the climate is in real trouble [1]. Has the 'treasure trove' become 'fools gold'?

The general public – or perhaps more accurately a section of it (small but growing) – is becoming more and more aware of weather trends and of the topic of global warming, although this awareness seems to be tinged with a worrying absence of concern. Extreme weather events are increasingly being reported in the media, and of course, hurricane Katrina which created havoc in the Caribbean and in the southern states of the USA in 2005, was perhaps the first really newsworthy story to nudge a few more people, over and above scientists and committed 'greens', to wonder 'Is there something in this global warming chatter?' 2005 was a year with an unusually high number of hurricanes, although 2003 did not do too badly either [1]. The summer of 2003 was apparently the hottest in Europe since 1500, but it was also a year of severe hurricanes. Causal connections between climate change, particularly global warming, and hurricanes have been a topic of much debate and not a few research studies. The growing consensus seems to be that, while our knowledge of the likely future evolution of the severity of hurricanes or tropical cyclones continues to remain an uncertain area of science, the correlation between the increased intensity of tropical cyclones and rising ocean temperatures is becoming increasingly difficult to refute [2]. It is worth noting that the exceptional weather of 2005 with the frequency and severity of its hurricanes has since been found to be in accord with the predicted trends. Hurricanes, and the very visual and graphic devastation which they cause, and the human interest stories which they spawn, yield good newspaper copy. As a consequence they have perhaps become the most effective climatic 'prongs' to hopefully prod awake the slumbering masses to at least consider the possibility that global warming is already here, and it could be potentially devastating!

Climatologists talk about a process of 'forcing' when quantifying the influence of atmospheric carbon on global warming. The Earth is naturally warmed by radiation from the sun. If you were to try to gather this solar heat over a square metre of the Earth's surface in daytime (obviously you would collect much more at the equator than at the poles) you would garner on average about enough heat to boil a three litre kettle of water. The sun produces, as one might anticipate, high energy radiation, which impinges on the Earth's atmosphere in the form of photons at light and higher frequencies. Some of these are scattered back out to space while the rest penetrate to the surface of the planet, with little absorption by the CO_2. On the other hand low energy radiation from the 'hot' Earth is at a much lower frequency and can be absorbed by CO_2 in the atmosphere. Man-made CO_2 is producing forcing (greenhouse warming) equivalent to 0.7% of the natural level; about enough solar power over a square metre of the Earth's surface to boil a table-spoon full of water. What this means is that a small fraction of radiation from the planet, which would normally propagate back out into space, is not permitted to do so by the enhanced CO_2 'blanket', and adds 0.7% to atmospheric and surface warming. This undoubtedly seems to be rather insignificant in relative terms, and consequently it is difficult not to ask: 'What is the problem?' The answer is that when scientists examine the ice core records, particularly at those periods in the distant past when there were pronounced atmospheric temperature increases of the order of 5°C, resulting in sea level rises of several metres, the CO_2 forcing is found to have been no more than three parts in one hundred (3%) of the direct solar warming. The additional greenhouse gases in the atmosphere which was, at that time, causing this forcing was, of course, 'natural' and due to methane and CO_2 leeching from the ground and the oceans because of enhanced solar warming, probably triggered by violent and protracted volcanic activity, at the Permian–Triassic extinction some 250 million years ago or a massive asteroid strike at the Cretaceous–Tertiary extinction 65 million years ago.

The Earth's orbit around the sun changes periodically from circular to elliptical in shape on about a 100,000 year cycle. At present the orbit is almost circular (eccentricity = 1) but it can have a value in the range 1.25–1.3. In this case the Earth can be carried much closer to the sun and it is additionally warmed during these excursions. The ice records of CO_2 and temperature faithfully echo these planetary movements, and these and other observations have been employed by scientist to compute a climatic sensitivity figure for CO_2. It suggests that the current man-made figure, which is producing 0.7% rise in warming over and above the natural background, will produce a mean temperature increase of just over 1.1°C. The average global temperature since pre-industrial times has risen by 0.8°C, so there is 0.3°C in the pipeline even if mankind maintains the status quo by cutting all new emissions. Maintaining the status quo is most certainly not what we are doing! It is estimated that growing carbon emissions will drive climatic forcing towards a very dangerous magnitude that will raise temperatures to 2–5°C above pre-industrial levels [3].

Worryingly, for future generations, it is estimated that at the end of 2007 there are still some 5000 Gigatons (five followed by twelve zeros!) of carbon remaining in the ground in the form of fossil fuels. If mankind does not 'awake' and continues to rely on fossil fuels to support prolifically energy wasteful lifestyles, then it seems highly likely that all 5000 Gigatons will 'go up in smoke'. In this case CO_2 in the atmosphere will rise four times above the pre-industrial level of 280 parts per million by volume, to say nothing of the fact that there will be little oxygen left. It is estimated that this level of CO_2 will drive climatic forcing well beyond 2°C. This is confirmed in a well researched article recently published in *Climatic Change* [4], where it is predicted, using results obtained from a range of very sophisticated climatic models, that global warming will be of the order of 5°C during the current millennium if atmospheric CO_2 rises four-fold. It is a change which dwarfs anything that we have seen in the last millennium. Conservative estimates suggest that the Greenland ice sheet is lost at about 2.7°C of local warming while the West Antarctica ice sheet could begin to disappear at 4°C. Consequently for largely coastal dwelling mankind, a quadrupling of CO_2 could be utterly devastating, with a possible total mean sea level rise predicted to be at least 15–20 feet, when thermal expansion of the oceans, is added to the effects of ice sheet disappearance and glacier loss.

In the summer of 2007, the North West Passage between Greenland and Canada was free of ice and open to shipping for the first time in recorded history. In 2008 the North East Passage opened for the first time. These events seemed to have little impact on the populace at large although, in many of the news reports, scientists were warning that a global warming 'tipping point' may have been breached. Why did this ecologically alarming event create much less impact than hurricane Katrina, which devastated New Orleans in August 2005? In my judgement the difference is explicable by one word – science. This news story contained scientific concepts such as positive feedback, albedo and tipping point. To grasp the significance of most versions of the story the reader was required to grapple with some science, and engaging with science represents a huge turn-off for an increasingly large majority of the population, not just in the UK but in many other parts of the world. It is my painful experience that to admit to being a scientist in social gatherings is to invite pariah status. To admit to being an electrical engineer is to invite a request to 'fix the washing machine'! Even in the most technologically advanced nations of the West, the vast bulk of their populations are, to all intents and purposes, scientifically illiterate. Unfortunately, current evidence suggests that this scientific ignorance is also endemic among our 'movers and shakers'. To find people who will proudly admit that, they are ignorant of how a computer or a mobile phone works, or they have not heard of Michael Faraday, yet would be embarrassed to admit they had read no works of Leo Tolstoy, or that they had not heard of William Shakespeare, is dismally commonplace. This lack of any scientific fluency among the vast multitude of the population must be hugely worrying for the 'concerned few' urgently seeking intelligent examination of the kind of energy and economic policy shifts, which may have some chance of properly addressing global warming.

It is immensely ironic that with the disappearance in the summer of 2007 of the Arctic sea ice, which is perhaps an early and significant symptom of global warming, countries bordering the Arctic Ocean are scrabbling to lay claim to the ocean bed. And why? In order, of course, to exploit the oil deposits that are predicted to exist there, despite the fact that their use will further degrade the ecological health of the planet.

1.2 Unstoppable 'Growth'

In addition to their ignorance of science, the imperviousness of most populations to the many global warming signals that have occurred recently is perhaps not surprising since, at this point in history, the political and business classes in Western society and increasingly in China and India, continue to be irrationally fixated on the 'market' and 'globalisation', although events in 2008 may be changing this. Our 'leaders' give no indication that they see global warming as a 'red light' to growth. Monetarism [5], introduced to the financial community by Friedman and others, and unwisely applied with gusto to the British economy by Margaret Thatcher in the 1980s, destroying the UK's engineering and manufacturing base, has become so established that it now seems to be viewed as an unchallengeable natural law of economics almost as if it were a 'law of nature'.

Yet this 'voodoo economics' of the Chicago School, as some have described it, is undermining the health of the planet, as we now know, at an alarming rate. Because of it, it is almost impossible for secular, democratically elected politicians to re-order their national economies, if they should happen to think that this might be necessary, with the reining back of growth, or perhaps even planned recession, as an aim. This is because of the nature of the global monetary system following state deregulation of the banks almost thirty years ago, and is exacerbated by the replacement of cash with technology. Quite simply, 'money' is issued as debt at interest. The system involves the creation by governments of only about 10% of the total money supply in the form of non-interest-bearing notes and coins, while the remaining 90%, over which they have little control, is created by the commercial banking system in the form of interest-bearing debt. At the instant when this debt is credited to each and every borrower, and there are so many the debt is huge, there is at that point no 'real money' being created by which the interest on the debt might be repaid. If the debts were immediately called in, the economic system would collapse, because there is insufficient real money to cover the debts. The solution to this dilemma is of course unstoppable economic growth, a run-away process, which requires, period by period, the creation of yet more credit, from which increasing arrears of interest can be paid. The encouragement which the system gives, to organisations and individuals with a propensity towards greed, acquisitiveness and financial irresponsibility, is quite disheartening. Adam Smith has often been attributed, possibly unfairly in the eyes of many, with extreme free market views, but in 'The Wealth of Nations', he has observed, rather presciently, given the banking collapse of 2008 triggered by financial recklessness, that the interests of the dealers and financiers are:

> always in some respect different from, and even opposite to, that of the public. The proposal of any new law or regulation of commerce which comes from that order (i.e. the dealers) ought always to be listened to with great precaution, and ought never to be adopted till after having been long and carefully examined, not only with the most scrupulous, but with the most suspicious attention. It comes from an order of men whose interest is never exactly the same with that of the public, who have generally an interest to deceive and even to oppress the public, and who accordingly have upon many occasions, both deceived and oppressed it.

This succinctly describes and illuminates the motivations of bankers, which have brought about the so called 2008 'credit crunch'. It also provides an explanation for the public disquiet at the 'bail outs', which have been proposed and introduced to rescue the troubled banking system at tax payers' expense. The dilemma has been aptly summarised by Neal Ascherson who has observed:

> The abject disasters of the credit crunch reveal something general about the age we live in. People no longer know what they are doing. There is too much information to grasp, too much technology and skill to master. The UK Northern Rock managers, who deserve little sympathy, had clearly lost any overall picture of what their liabilities were, or of how shoogly their whole structure had become [5].

The same seems to have been true of managers at Bear Stearns, Fannie Mae, Freddie Mac and Indy Mac in the USA.

Unremitting growth, then, is essential to the current global economic/financial system. Yet from an engineering perspective, it seems to me that a system that is endlessly expanding on a finite Earth, cannot help but contravene two very basic laws of physical systems, namely the first and second laws of thermodynamics, which preach conservation and the inter-dependence of natural processes, and consequently that if contravened, there must be a 'price to pay'. Of course, economists tend to obscure possible difficulties with continuous growth and the 'endless' supply of materials required to feed it, by disingenuously talking about resources when they mean reserves. While new coal, oil, gas and other mineral supplies continue to be discovered, reserves are seemingly unlimited and so the unsustainable features of endless growth can be disguised. This will not always be the case. When there is no more coal, oil, gas or minerals to be found, reserves and resources will become synonymous and then the limits will be stark.

From the start of the industrial revolution, which has primed, and provided the engine of economic growth, the source of the energy and materials fuelling the industrial dynamo has been very largely of the non-renewable variety, and this fuel is very obviously a finite resource. On the other hand, monetarism or unfettered capitalism, which is 'fuelled' by ever expanding and seemingly unlimited money supply through easy credit and borrowing, requires an unlimited supply of non-renewables (fossil fuels, metals, minerals), which must be exploited at an ever increasing rate, if global inflation is to be avoided. The economic system is now 'hitting planetary buffers' – something must give? Of course this danger was predicted some 35 years ago in the seminal work *Limits to Growth* (LTG) published in 1972, where the authors made the following statement:

> If the present growth trends in world population, industrialisation, pollution, food production, and resource depletion continue unchanged, the limits to growth on this planet will be reached sometime within the next 100 years. The most probable result will be a rather sudden and uncontrolled decline in both population and industrial capacity [7].

In *Slow Reckoning* [8], a powerful analysis of the North–South divide in an ecologically challenged world, Tom Athanasiou makes the following pertinent comment in referring to the views of 'greens'.

> Though they seldom name (industrial) society as "capitalist", their insistence that "growth" must end is the core of the green challenge to capitalism, and though it is often ignored, it is never effectively refuted. Capitalist economies must expand, but the ecosystem that is their host is finite by nature. It cannot tolerate the indefinite growth of any human economy, least of all one as blindly dynamic as modern capitalism.

It seems pretty clear now, having perhaps observed in 2008 the first twitches of global capitalism's death throes, that by 2030 worldwide recession will be well under way – but we also know now that the misery of recession will be exacerbated by growing climate unpredictability and ferociousness, which will make life additionally difficult and precarious for many. This is also acknowledged in the 30 year update to LTG [9] where the following observation is made: 'humanity is already in unsustainable territory. But the general awareness of this predicament is hopelessly limited'. A CNN news report of the 25th June 2008 contained the following announcement, confirming this lack of awareness: 'World energy use is expected to surge 50% from 2005 to 2030, largely due to an expanding population and rapid economic growth, according to a government report. Without any new laws restricting greenhouse gases, carbon dioxide emissions will see a similar jump, the Energy Information Administration [10] said in its annual report on global energy markets'.

It is sad to relate that a well respected science journal, namely *Scientific American*, was strongly expressing a 'growth is sacrosanct' mindset as late as 2006 [11]. In a Special Issue devoted to 'Energy's Future Beyond Carbon' with the subtitle 'How to Power the Economy and Still Fight Global Warming' the nine articles were focused, in one way or another, on offering continued economic growth while 'solving' global warming. The content of this special edition is extensively and trenchantly reviewed by A. A. Bartlett in the *Physics Teacher* [12]. In the first article, 'A Climate Repair Manual', global warming is acknowledged to be a major problem. The article concedes that: 'Preventing the transformation of the Earth's atmosphere from greenhouse to unconstrained hothouse represents arguably the most imposing scientific and technical challenge that humanity has ever faced'. It also suggests that 'Climate change compels a massive restructuring of the world's energy economy. The slim hope, that atmospheric carbon can be kept below 500 parts per million, hinges on aggressive programmes of energy efficiency, instituted by national governments.' However, the 'massive restructuring' alluded to, is in global economic terms, of the minimally disruptive, market friendly, techno-fix variety. Athanasiou [8] puts it this way:

> For now "the era of procrastination, of half measures, of soothing and baffling expedients" continues, and swells of talk assure us that only what is minimally disruptive is actually necessary. And then the circle is closed – since little is done to face the situation, that situation must not be so serious after all. The alternative conclusion, that we are drifting almost unconsciously into a mounting crisis, is not admissible.

Even otherwise sensible contributions to the debate, for example Zero Carbon Britain [13], tend to suggest that zero carbon emissions by 2030 can be achieved through techno-fixes and market economics using what they describe as tradable energy quotas (TEQs). In a long established civilised democracy like Britain there is a remote possibility that this kind of scheme would be accepted and responsibly administered. However, in the context of a relatively uncontrolled global market, it seems to me that TEQs are open to exploitation and to the growth of business and finance inspired profit-making scams. What is required is the rationing without the trading. Common sense suggests that any market driven mechanism for forcing down carbon emissions is probably doomed to ignominious failure, as the 2003 Kyoto Protocol, based on carbon trading, has already demonstrated (see Sect. 1.4). If one believes the evidence, it is difficult not to conclude that we really need to leave all of the remaining fossil fuels in the ground, where it is doing no harm to the planet. No market process will result in such an outcome.

A secondary but very important driver of economic growth is, of course, the swelling world population, but virtually nowhere in the mass media, or in political discourse, is this acknowledged in 2008. To quote LTG once again: 'For generations both population and capital growth were classified as an unmitigated good. On a lightly populated planet with abundant resources there were good reasons for the positive valuation. Now with an ever clearer understanding of ecological limits, it can be tempting to classify all growth as bad'. The biologist and broadcaster Aubrey Manning, based at Edinburgh University, has put the same point in this no nonsense way: 'Population growth is linked to economic growth. People talk about sustainable development all the time. What they usually mean is sustainable growth, which is by definition not sustainable'. Some significant fraction of the observed global warming is now widely acknowledged to be caused by the release of greenhouse gases from the burning of fossil fuels. In a global market, as the size of the world population expands, clearly the rate of burning of fossil fuels increases and this can be expected to drive up the rate of rise of global average temperatures. Most contributors to the warming debate appear not to recognise, or prefer not to acknowledge, that the size of the Earth's population, economic growth, and the expansion of 'western' lifestyles into all parts of the world, is a major factor in determining the rate of release of greenhouse gases. This consequence is simply an obvious manifestation of the laws of thermodynamics in action. That the special issue of Scientific American mentioned above, which is supposedly devoted to reducing global warming, almost completely ignores profligate energy usage and population size, is quite mystifying. Instead it directs our attention towards a range of potentially profitable, technology rooted schemes, which will underpin and reinforce continued growth and support rising population. It should be noted that there is a 'Denial Industry', particularly in the USA, which works very hard and assiduously to perpetuate and encourage views of this ilk [14].

Again for those who are inclined to believe the scientific evidence a couple of rather apt clichés come to mind at this point in time. These suggest that mankind is either about to 'hit the buffers' or that for dwellers on planet Earth the 'chickens are coming home to roost' – or more crudely that 'the shit is about to hit the fan' in the guise of irreversible global warming unless current trends can be arrested. So could the widespread adoption of presently advocated market driven technofixes, aimed at expanding energy supply from renewable resources, be enough to do this? I shall try to address this question in the following chapters.

1.3 Eye of the Beholder

Scotland, a small ancient nation to the north of England, and a part of the United Kingdom, generates a relatively high proportion (13%) of its electricity from renewable sources, predominantly taking the form of hydro-electric generation. However, most Scots, and many of the visitors to Scotland, and to the remote places where the hydro-schemes have been sited, would hardly complain that the scenery is tarnished by their presence. Some might think that a few of the Scottish dams actually add to the grandeur of their location. On the other hand, a zealot for wilderness might see only man-made artefacts that are 'polluting the landscape', but this would be an extreme view. The concept of 'wilderness' is becoming increasingly difficult to promote in today's world, which has become highly sculpted and modified by mankind, in order to support a population that has rapidly outgrown the ability of the planet to sustain it naturally. Wilderness is where modern human beings have never been and where their presence on the planet is not apparent. Where on Earth is that! When one sees photographs of remote mountains, remote islands and even very remote, seemingly pristine Antarctica, showing evidence of contamination originating from human activity, it is clear that humanity's flawed stewardship of the planet has resulted in there being really nowhere left where it is possible to view truly unsullied landscape or seascape. James Lovelock, the renowned originator of the Gaia hypothesis, who was a young man in the 1940s, has cogently opined that:

> Even in my lifetime, the world has shrunk from one that was vast enough to make exploration an adventure and included many distant places where no one had ever trod. Now it has become an almost endless city embedded in an intensive but tame and predictable agriculture. Soon it may revert to a great wilderness again.

In making the above observations it is difficult, as a scientist, not to be reminded of a rather famous experiment created by John B. Calhoun [15]. It has since been widely referred to as the mouse universe. In July of 1968 four pairs of mice were introduced into this Utopian universe – at least for mice. The universe was a 3 m square metal pen with 1.35 m high sides. Each side had four groups of four vertical, wire mesh 'tunnels'. The 'tunnels' gave access to nesting boxes, food hoppers, and water dispensers. There was no shortage of food or water or nesting material. There were no predators. The only adversity was the limit on space.

> Initially the population grew rapidly, doubling every 55 days. The population reached 620 by day 315 after which the population growth dropped markedly. The last surviving birth was on day 600. This period between day 315 and day 600 saw a breakdown in social structure and in normal social behaviour. Among the aberrations in behaviour were the following: expulsion of young before weaning was complete, wounding of young, inability of dominant males to maintain the defence of their territory and females, aggressive behaviour of females, passivity of non-dominant males with increased attacks on each other which were not defended against. After day 600 the social breakdown continued and the population declined toward extinction [15].

The conclusions drawn from this experiment were that when all available space is taken and all social roles filled, competition and the stresses experienced by the individuals involved will result in a total breakdown in complex social behaviours, a despoiling of the habitat, ultimately resulting in the demise of the population. Dr. Calhoun believed that his research provided clues to the future of mankind as well as ways to avoid a looming disaster. One would like to think that there should be no parallels between mice and men, but the evidence is not encouraging. Of course, Rabbie Burns, if he were alive today, with his knowledge of the nature of the 'timorous beastie', would not be surprised, either at the results of the experiment or at Dr. Calhoun's inferences. Rabbie Burns is just possibly the most influential Scot who has ever walked on the surface of the planet after James Clerk Maxwell.

While wilderness may no longer exist we should of course be concerned to preserve significant areas of the planet where nature can be given 'free reign'. Balancing 'nature' and human 'progress' has been a difficult problem for human society since the industrial revolution and it will greatly increase in a world with a population approaching 10 billion, dependent wholly on renewable resources. If, as we have seen, significant levels of electrical power can be extracted from reservoirs and dams, without blotting the landscape, when these are sensitively located, how much is this likely to be true of other renewable resource collectors. Hydro-electric schemes are a good example since these are well established and exist in sizeable numbers in several countries, such as Canada, Norway and Sweden, yet in their building, the evidence suggests that local populations were not often outraged by any perceived environmental damage, although others may have been intensely distressed by losing flooded homes. It is also appropriate to note that some of the images emanating from China and India, are quite disquieting, demonstrating that even today hydro-electric power developments are not necessarily friendly to the local environment, particularly at the civil engineering stage. But it seems likely that once they are 'bedded down' and operational, that they will gradually merge into the landscape much as long established hydro-power stations have done. Most of the Scottish hydroelectric schemes – there are a lot of them – are impressively in character with the scenery, and it is difficult to imagine that they could give offence to walkers or climbers seeking to enjoy the rugged Scottish landscape. The environmental impact of hydro-schemes like these is not negligible of course, but neither is it gross, unlike unsympathetically routed major roads and motorways, the careless siting of visually unappealing petrol/gas stations, or of conspicuous agri-business warehouses and sheds, to name but a few human constructs, which litter the countryside. Nevertheless, it seems pertinent to ask to what extent this experience of inoffensive and uncontroversial hydro-schemes remains true in other parts of the planet?

Recorded images of the reservoirs and dams of the world, and travelogues, which report the impressions of professional itinerants, are not difficult to track down. Extensive and wide ranging picture galleries are to be found on the web. Photographers, presumably with a 'good eye' for scenically appealing views, seem to find that hydro-electric dams are worthy of their attention. It is probably fair to say that the best dams have a rugged beauty and a grandeur which makes them aesthetically appealing, and in viewing them it is possible also to see impressive engineering (Fig. 1.1), which has enabled a large water storage and electrical power generation problem, to be solved with elegance. Of course with images one has to be cautious these days, since 'doctoring' is easy, but the evidence seems to

Fig. 1.1 The impressive Hoover Dam straddling the Black Canyon of the Colorado River in Arizona. The scale can be gauged from the vehicles and cabins on the cliff ledge to the right of the dam and in the forefront of the photograph

be that the majority of hydro-systems around the globe are by no means scarring the landscape.

Visually dams are not unlike bridges. The best are stunning, while most are commanding, because they represent raw man-made strength resisting the power of nature, but expressed in elegant engineering language. A testament to this statement is the fact that the Itaipu Dam, between Brazil and Paraguay, is listed as one of the wonders of the modern world. They are structures which are designed for a very specific purpose, perhaps like castles in a former age, and that purpose informs their design. It seems not unreasonable to suggest, therefore, that hydro-electric schemes, once built, in addition to being ecologically benign, contribute little in the way of visual pollution to the natural environment – a growing feature of modern life. Unfortunately, in the past forty to fifty years, planning authorities at the behest of politicians, who have been prone to making poor choices to accommodate swelling populations, and burgeoning car ownership, have succeeded in furnishing the industrialised world with a plethora of rather depressing towns and cities. These dystopias are generally a disagreeable mixture of urban dereliction and sub-urban sprawl criss-crossed with ugly streets that have been subordinated to the car and other road vehicles, to the obvious detriment of all else. Furthermore, the intervening countryside, or what is left of it, is degraded by vast motorways systems, interspersed with drab motorway service stations, grim out-of-town supermarkets, sprawling industrial estates, belching refineries, and dismal airports. It would not be difficult to add to this list. Human beings, it seems, are generally much better at diminishing the natural landscape, than enhancing it, with their buildings and artefacts. Of course there are a few exceptions to this human predilection for scarring the countryside. Ironically these, because they have become visual treasures, are themselves being spoilt by unsustainable visitor numbers. The relevance of these jottings is this; as a species, we seem to be doing pretty well at degrading most of the visually uplifting vistas on Earth, that still remain to be enjoyed. Consequently, complaints about the deleterious impact of emerging renewable power stations, such as wind farms, are hard to take seriously, particularly since these 'intrusive objects' could help to preserve the ecological health of the planet.

In fact, the visual and environmental disturbances likely to be incurred by many sustainable power stations, such as those employing wave, or tidal, or geothermal energy sources, are not going to be of significant concern to the public, since the infrastructure, as we shall see, is of limited size (like oil wells or coal mines), and there is no reason for the associated generating plants to be other than sparsely distributed over the surface of the Earth. On the other hand wind farms and solar power stations are potentially vast, for reasons which will be explained in Chap. 3. In some parts of the world renewable power systems, but wind farms in particular, are being introduced in a piece-meal, apparently uncoordinated fashion, which raises questions as to their effectiveness. Consequently, despite the atmospheric advantages accruing from their adoption, it is inevitable that some special interest group with profound concerns about the destruction of treasured scenery and natural landscape will raise objections to their construction. Obviously the need, to balance the visual impairment and the possible ecological harm to the natural environment, which technology can cause, with the demands of the growing economies of the industrialised world, is not new. In fact, the scales have usually been weighted heavily in favour of economic advancement.

Technology for a sustainable future is perhaps a bit different from developments in the past, which have generated much anger and heated debate among environmentalists – in some cases with good reason. It is a pity CO_2 and other greenhouse gases are not slightly opaque to light, like an urban smog of the 1950s, but maybe not so dense. If environmental campaigners against wind farms and solar farms were able to see their precious landscape only indistinctly through a blurring haze of CO_2 gas, they would soon accede to the need for extensive 'forests' of wind and solar collectors. Mind you not everyone dislikes these forests. The inestimable newspaper columnist, Ian Bell puts it this way [16]: 'As blots on the landscape go, wind farms are not the worst. I would really like to pretend to think differently, but I don't, and I can't. Beyond the pale I may be, but to my eye these things are pretty enough, in a good light. So there'. In the Scottish paper, the Herald, of the 27th July 2008, in the letters page, David Roche remarks: 'The plains of Denmark and north Germany have massive wind farms which provide spectacular visual interest in a flat landscape'. It is difficult not to conclude, from all this, that any environmental damage brought about by the emerging infrastructure associated with an electrical supply industry built around renewable sources of power, is unlikely, at this point in time, to add much to the degradation that has already been perpetrated on the planet by mankind, during the era of fossil fuels.

1.4 Techno-fix Junkies

The obvious, but uninformed, response to the 'warming' dilemma, which is being strongly pushed by the financial community and by our political leaders, is a switch away from our reliance on fossil fuels through the agency of a market led expansion of power generation from so called renewables, such as wind power, wave power, tidal power, hydro-electric power, solar power and geo-thermal power. Nuclear power is usually included in the mix but it is not really renewable unless scientists can crack the nuclear fusion riddle, and that seems to be unlikely in the foreseeable future. Also, biomass has been excluded here because it is not really a viable solution using land based crops, if the swelling population of the planet continues to demand to be fed [17]. Europe has already announced (in 2008) cut-backs in recent targets for the percentage of vehicle fuels which should comprise bio-fuel. Seaweed cultivation has recently been mooted as a source of bio-mass but it is highly unlikely to be providing serious quantities of fuel by 2030.

Ingenious, but fanciful, notions of alleviating global warming by reflecting the suns rays back into space, while probably devised for the best of reasons, nevertheless represent, quite frankly, rather inappropriate and misguided applications of geo-engineering. In this geo-engineering category I would place the following: seeding space with 20 trillion metre-sized optically reflective mirrors [18]; seeding stratocumulus clouds over the oceans to make them whiter by spraying huge volumes of sea water into the upper atmosphere [19]; introducing sulphate aerosols into the stratosphere to reflect sunlight using high flying aircraft [20]. For mankind to pursue the application of any of these, and others, would be not unlike the crew of a ship on the high seas, which is listing dangerously due to a shifting cargo, and instead of correcting the problem by applying all their effort into restoring the cargo to its original position, they choose to try to counteract the list by following the much more risky course of attaching novel list-compensating bow planes to the keel of the ship. Needless to say, some advocated techno-fixes are rather too risky to be treated seriously. As Lovelock [21] has observed 'geo-engineering schemes could create new problems, which would require a new fix – potentially trapping Earth into a cycle of problem and solution from which there was no escape'.

In a late night programme on BBC television (13/3/08) entitled, 'This Week', hosted by the arch right-wing broadcaster Andrew Neil, the regular political commentators Michael Portillo (a former UK defence secretary), and Diane Abbott (UK Member of Parliament), were confronted by journalist Rosie Boycott about global warming and mankind's energy profligacy, which was obviously a topic of great concern to her. She wanted to know what politicians really thought about the issue given that Alistair Darling's first budget, the previous day, had been predictably anaemic on global warming measures. Portillo summarised pretty well the attitude of the political classes when he said, 'First, I don't think the problem (of global warming) is as significant as people (green campaigners) like Rosie think it is. Secondly, they (politicians) don't think people want to address their behaviour.

All sorts of votes are there to be lost (if they are made to). Thirdly, they probably think the problem is solvable not by people adjusting their behaviour, but by (moving to) new technology – nuclear (power generation) and hydrogen powered cars'. Neil then suggests that this means 'the solutions can be painless'? Portillo agrees. With this kind of response from a reasonably intelligent politician, who comes across as possessing a good sense of how 'the political wind is blowing', the prospect for real action in the near future is really rather grim.

Considered views on the issues raised by global warming can be found in the literature if you look hard enough. Readers are referred particularly to MacCracken [3], Monbiot [14], Romm [22, 23], Tickell [24], and Flannery [25]. MacCracken, in particular, provides copious information and detail on the physics, and the mechanisms, causing the increase in CO_2 in the atmosphere, with explanations and evidence of the linkages to global warming. All broach the issue of providing techno-fixes to supply future energy needs, although Monbiot concentrates on UK requirements. But a coherent solution seems always to flounder on how it is paid for, when economic growth is sacrosanct, and the 'global market' has to be retained as the only viable mechanism for changing human behaviour away from reliance on fossil fuels. Tickell puts it this way: 'Energy efficiency and low carbon developments are laudable objectives so long as we understand what they are for – to enable continued economic growth and human welfare gains under a greenhouse emissions cap, and so making the cap consistent with economic and political imperatives'. The impression given, which is surely not intended, is that these imperatives are more important than the health of the planet! Population growth is given some space by Tickell, but is otherwise hardly mentioned as an issue. The message from this more 'serious press' is that anthropogenic global warming is real and measurable and that it can no longer be ignored. A transition from fossil fuels to renewables is inevitable – sooner rather than later. The financial and social costs of making it happen are huge, possibly on the scale of waging a world war. But this is for others to ponder.

Unfortunately, the electorates in western democracies, despite the growing numbers of cautioning voices, are mostly being promised, that 'new' sources of power will provide the needs of unremitting growth, and lifestyle changes will not have to be forced upon unwilling populations. Many committed 'greens' and concerned scientist would view this incoherent embracing of 'renewables' as a short term technical fix, which, reluctantly, has to be countenanced at this early stage in the response to global warming, because of the huge inertia to meaningful change in human societies. Recently, even Professor James Lovelock [21], the author of *The Revenge of Gaia*, and a techno-fix sceptic, has expressed reluctant approval, to the dismay of 'greens', for the introduction of new nuclear power stations into the UK because he has become aware that any productive discussion at influential levels, of the real solutions that are required, is remote. Effective and lasting solutions are too unpalatable to be addressed by politicians seeking a democratic mandate, since in addition to technology, they are likely to involve drastic cuts in energy usage by mankind as a whole (planned recession), together with serious reductions in global population levels within the current century. Unfortunately the 'over egging' by the 'market' of technical fixes, of seemingly unlimited capacity, and the consequential reassurance they offer to the layman that science/engineering on its own can solve our dilemma, has the undesirable effect of convincing the technically ignorant, political class, the financial community and the business community that 'business as usual' is possible. That is, that mankind can continue with its energy profligate and wasteful lifestyles into the foreseeable future. To this Lovelock is quoted as saying 'that carrying on with "business as usual" would probably kill most of us this century'.

This 'business-as-usual' mind set is displayed clearly in the much quoted foreword to the report [26] to the G8 summit written by Tony Blair, the former UK prime minister. In it he writes: 'If we are not radical enough in altering the nature of *economic growth* (my italics) we will not avoid potential catastrophe to the climate'. In other words, whatever we do to mitigate climate change, cannot harm growth! His solution is the extremely costly nuclear techno-fix, presumably not realising that economically exploitable uranium ore, would soon run out if there were a very substantial rise in nuclear electricity generation. Even at present rates of extraction it will run out in 85 years. There is no mention of measures to address population growth, to curb the market and rampant consumerism, to curtail unsustainable air travel or to introduce measures to drastically reduce reliance on road vehicles. His weak grasp of the seriousness of global warming is highlighted by the following confused utterance:

> We are not assisted by the fact that many of the figures used are open to intense debate as our knowledge increases. For example, we talk of a 25–40 percent cut by 2020. But, to state the obvious, 25 is a lot different from 40 percent. Some will say that to have a reasonable chance of constraining warming to approximately 2°C, we need greenhouse gas concentration to peak at 500 parts per million by volume (ppmv); some 450 ppmv; some even less. Some insist that 2020 is the latest peaking moment we can permit, beyond which damage to the climate will become irreversible; some, though generally not in the scientific community, say 2025 or even 2030 may be permissible. [26]

The global warming process and its consequences at each level of temperature rise, have been powerfully and graphically described by several contributors to the global warming debate [3, 14, 22, 23, 24, 25]. There is little room for dubiety, for anyone predisposed towards rationality. For example, Monbiot [14] expresses the view that mankind still has a window of opportunity to forestall runaway warming by doing all we can to stabilise atmospheric carbon at a level that ensures that the planet does not reach the 2°C 'tipping point'. At the present rate of increase it is predicted to occur in about 2030 when the global average temperature will have risen by about 1.4°C from where we are now (2007). As Monbiot says, 'Two degrees is important because it is widely recognised by climate scientists as the critical threshold'. But we must start making really significant reductions in the rate at which we are burning fossil fuels now – not in 2020 or 2030 as Blair seems to be suggesting!

A UK Meteorological Office conference paper [27] published in 2005 predicts that by 2030 the Earth atmosphere's capacity to absorb man-made CO_2 will have reduced to 2.7 billion tons a year from the current level of 4 billion tons. What this means is that by 2030 mankind can pump no more than 2.7 billion tons a year of CO_2 into the atmosphere if we wish to ensure that the concentration of CO_2 remains stabilised at a level (440 parts per million by volume – ppmv) which is consistent with not breeching the 2°C temperature rise bench mark. More recent evidence [24] suggests that 440 ppmv may be too high, and that 300–350 ppmv will have to be achieved by 2050. The problem is that the world as a whole currently pumps three times more than is prudent, namely 8.4 billion tons/year, into the biosphere [3] (and this figure is rising not falling), most of it by Western countries, with China and India making every effort to catch up. The danger in following this course is starkly illuminated in this quotation from Lovelock [21] in relation to a prehistoric period of mass extinction:

> The best known hothouse happened 55 million years ago at the beginning of the Eocene period. In that event between one and two teratonnes (2×10^{12} tonnes) of carbon dioxide were released into the air by a geological accident. [….] Putting this much CO_2 into the atmosphere caused the temperature of the temperate and Arctic regions to rise 8°C and of the tropics 5°C, and it took 200,000 years for the conditions to return to their previous state. In the 20th century we injected about half that amount of CO_2 and we and the Earth itself are soon likely to further release more than a teratonne of CO_2. [24]

Monbiot has done the sums and estimates that by 2030 when the global population will be ~8.5 billion, equitable rationing will demand that a maximum allowable emission rate of 0.33 tons/year for each person on the planet is somehow introduced. In prosperous countries, such as the USA, Canada, European nations, Australia, this means that an average cut in CO_2 production of the order of 90–95% (on 2005 levels) will be required by then. This percentage figure is massively in excess of anything which has been agreed to by the countries that have signed up to the 1997 Kyoto Protocol. The protocol came into force on the 16th February 2005 and it commits 36 developed countries plus the European Union to meet specified reduction targets by 2012. The agreed amount varies from country to country but is of the order of a risible 5% cut in total carbon emissions by the target date. Even at this low target level governments have chosen to bend carbon trading rules so that Kyoto targets will not be met [24]. Tickell also suggests that 'Indeed the funds from the sale of carbon credits appear in some instances to be financing accelerated industrial development – and actually increasing emissions'. Yet Blair, in the 2008 G8 report [25] talks about trying to gain consensus on a pathetically inadequate 50% reduction in emissions by 2050! A global reduction of the order of 90% by 2030 would appear to be in the realms of fantasy, particularly when the only solution which is on the political and business agenda is of the market friendly techno-fix variety. The market friendliness of Kyoto is underlined in a quotation from UK Prime Minister, Gordon Brown in a speech in 2007:

> Built on the foundations of the EU Emissions Trading Scheme, with the City of London its centre, the global market is already worth 20 billion euros a year, but it could be worth

> 20 times that by 2030. And that is why we want the 2012 agreement, the post-2012 agreement, to include a binding emissions cap for all developed countries, for only hard caps can create the framework necessary for the global carbon market to flourish. [24]

In other words the 'flourishing' of the global carbon market is much more important than curtailment of carbon emissions. We now have plenty of evidence to conclude that this market does not seem to be helping the planet.

The dangers of endless procrastination, at governmental level, are placed firmly in the spot-light in a report from the New Economics Foundation [28], which expresses the situation quite uncompromisingly with the following observation:

> We calculate that 100 months from 1 August 2008, atmospheric concentrations of greenhouse gases will begin to exceed a point whereby it is no longer likely we will be able to avert potentially irreversible climate change. 'Likely' in this context refers to the definition of risk used by the Intergovernmental Panel on Climate Change (IPCC) to mean that, at that particular level of greenhouse gas concentration (400 ppmv), there is only a 66–90% chance of global average surface temperatures stabilising at 2°C above pre-industrial levels. Once this concentration is exceeded, it becomes more and more likely that we will overshoot a 2°C level of warming. [28]

Notwithstanding the fecklessness of politicians on this issue, it is rather intriguing to observe how the global warming debate, at least where it has been intelligently joined, has firmly gravitated towards, and become focused on, technical solutions based on so called 'renewables', which place heavy reliance on electrical generation and transmission. Despite the 'rights or wrongs' of the global warming debate, the process of switching to renewables will have to be engaged eventually as fossil fuels become exhausted. In so far as this impression of an electricity dominated future is valid, it seems that it is relevant to attempt to provide a view of the transition issue that emphasises and focuses upon the engineering questions. What appears to be missing, so far, is a considered description and evaluation of the technology that might be capable of delivering renewable electrical power in the relevant time scale, plus an assessment of how far these proposed electro-technical developments can advance the search for a solution to the environmental dilemma, or if you prefer the crisis of disappearing fossil fuels. Obviously they are linked, and both have to be addressed. A further aim will be to collect and evaluate the evidence of real technological progress, if any, that is being made to wean mankind off fossil fuels, and to determine how far the currently incoherent 'dash to renewables' can go towards providing a sustainable future with advanced living standards for more than 10 billion people. As presently enunciated and propounded, current market led plans to arrest global warming appear to be little more than 'green-washing', and seem unlikely to achieve even the most modest of sustainability goals. To perform this evaluation the accepted scientific approach of reducing the parameters of a complex problem to a manageable level has been followed without, hopefully, losing its essence. This has been done by largely suppressing those parameters that measure political, economic, ecological and environmental concerns since, although they are obviously important, they are peripheral to the need to develop an appreciation and a proper understanding of the purely engineering implications of the demise of fossil fuels and the transition to sustainable sources of power, should it come to pass.

The question that is still before us is this: can an impending global warming disaster be averted by moving to renewably resourced electrical power? What can be achieved by piece-meal techno-fixes? If we persist with the current, market led, exploitation policies can a sufficient proportion of the global demand for energy by 2030 be supplied from renewables, which would enable the industrial world to meet even the most modest emission reduction targets? In the longer term, can integrated electrical power supply systems based on renewables be constructed to both replace fossil fuels and accommodate the energy demands of modern societies? What sort of technology would be involved in implementing such systems? Do we have the technology? These question are addressed in Chaps. 3 and 4.

1.5 Dearth of Engineers

A paradigm shift in the energy supply infrastructure for the planet is being hesitantly postulated. It will entail, if implemented properly, the abandonment by mankind of fossil fuels in favour of renewable energy sources to generate electricity for all our energy needs. This is a potentially massive undertaking that cannot possibly be implemented without huge engineering effort. We are, in effect, going to have to create an industrial goliath, of similar proportions to the current automobile and aerospace industries combined, to produce renewable infrastructure at the pace required. From an engineering perspective, it is difficult not to ask the following question: 'Where are the professional engineers going to come from, given that there has been a serious dwindling of recruitment into engineering and science courses in our colleges and universities for the past 20 to 30 years?' This conundrum is especially apposite in relation to the older industrialised nations in North America and Europe, and for nations such as Japan, Australia and New Zealand. To avoid an engineering skills dearth in these parts of the world, it is going to be necessary to massively expand education provision in an unprecedented way, which will ensure that colleges 'roll out', in sufficient numbers, the engineers, scientists and technicians that are going to be in demand between now and 2030, to propel what is nothing less than a renewables revolution – if it is initiated. It has been suggested that the energy industry 'tanker' is proving to be ponderously slow to turn towards renewables. However, this geriatric gait could possibly appear more like a foaming speed boat by comparison with the 'leviathan' of the education sector, a sector which is notoriously slow to change. As a former 'insider', it seems to me that if the call comes for more science and engineering graduates it is almost inevitable that the education sector's response will be lethargic to the point of immobility.

The problem for the education sector in the 'old' industrialised world is a disinterest in, and a lack of enthusiasm for, 'technology', particularly among the young, but also among not a few school teachers with weak science and mathematics backgrounds. In the UK, educational bias against engineering is not new. Even fifty years ago it was my experience as someone wishing to pursue a career in engineering, having qualified to enter university, that the available advice from teachers and others was distinctly unsupportive. This kind of reaction was not uncommon then and the indications are that it is even worse now, particularly in schools where the most able recruits are to be found. Whereas then, the advice was to study pure science or even the arts rather than engineering, now I suspect it is business, finance, and the law! Furthermore, it is becoming clear that a peculiar notion seems to have germinated in schools that 'education should be fun' and that it should be more about 'self-improvement and self-knowledge', than about 'understanding the physical world'. Sadly, even in our universities the idea is taking root that all knowledge is ephemeral and that it is skills which should be nurtured, since skills are forever. Such an ethos does not create engineers!

Science and mathematics teaching requires that students should be prodded, cajoled and encouraged to grapple with ideas and concepts that are often counter-intuitive, and which demand considerable mental effort, before understanding is secured. The joy of the 'eureka' moment, which makes the intellectual effort all worthwhile, is being experienced sadly, by fewer and fewer students. In the secondary schools, where students make decisions about the university courses they will pursue, there is an acknowledged shortage of teachers in mathematics and physics, the essential precursors of undergraduate engineering studies [29]. My experience of many years teaching undergraduates in electrical engineering science, is that today, few students entering universities in the UK to study science or engineering have understood, or accept, the need to 'sweat a little' in order to gain mastery of an intellectually difficult topic. Anecdotal evidence suggests that diminishing technical skills among university entrants is also an issue in many other countries. The problem is that the pupils, from whom the required new engineers will have to come in very large numbers over the next twenty or so years to advance the 'renewables revolution', are already in an education system that values self-expression over numeracy. Consequently few are likely to gravitate towards engineering without massive incentives.

The drift away from engineering and science has also been exacerbated by the lack of role models in the medium that arguably has most influence on the thinking and attitudes of youngsters – namely television. This is not just a UK phenomenon. Aspiring medical doctors, lawyers, financiers, and business men have plenty of programmes that extol their roles in society, but you will look hard to find a programme that depicts an engineer as other than a repair man. Not that there is anything wrong with repair men (or women) but I doubt if even their doting mothers would describe them as professional engineers or fully trained technicians.

A comprehensive UK report [30] in 2002 stated the following: 'Engineering has an image problem resulting in a short fall (in 2001) of 21,000 graduates. An important message engineering educators need to get across is the far wider applications of their subject, raising awareness and understanding of engineering'. The report notes that, at the time of release, the basic output of engineers was effectively stagnating. Between 1994 and 2004 the number of students embarking on engineering degrees in UK universities remained static at 24,500 each year even though total university admissions rose by 40% over the same period. Further, after completing their studies less than half of UK engineering graduates subsequently choose to enter the profession [30]. The statistics have got worse since then, and the raw statistics do not 'paint the full picture'. The kind of electrical engineers we will be seeking, to advance the putative 'renewables revolution', are those with competency in electrical power and high voltage engineering. Unfortunately these topics are very unfashionable even among students studying electrical engineering, most of whom would rather study computer and communications orientated subjects, such as digital circuits, integrated circuits, signal processing, image processing and software engineering. Electrical power engineering courses are in danger of disappearing from many electrical engineering degrees in the UK, and there is little doubt this situation is being replicated in universities throughout the industrialised west.

The declining of engineering subjects in schools is growing, not just in North America and the UK, but in schools in Europe, Japan and Australia. International developments elsewhere make the implications of this situation not a little disquieting. Mature economies, such as that of the UK, must now compete with those of rapidly developing countries such as the BRIC nations – Brazil, Russia, India and China. On current projections the combined gross domestic products of the BRIC nations are set to overhaul those of the G6 countries (US, UK, Germany, Japan, France and Italy) by the year 2040. Furthermore the BRIC nations are producing record numbers of graduate engineers (but mainly civil and mechanical) to build the infrastructure of their rapidly expanding economies: powered, of course by coal and oil. In China and India alone, the most conservative estimates suggest that around half a million engineers now graduate each year [31]. Many of these engineers will hopefully gravitate from fossil fuel powered developments towards the task of creating a renewables based infrastructure. The potential for the BRIC block of nations to out-muscle the 'old' industrialised world in harnessing the technologies of the future is high, unless very large numbers of 'new' engineers can be plucked from the colleges of the G6 nations soon, and not by resorting to 'creative accounting'! The statistics are not encouraging for the industrialised west. In an Engineering Council survey in 2000 of the engineering profession [32] it is observed that: 'In Germany, over the period 1991–1996 the numbers of students entering science and engineering dropped by a startling 50%. In the USA, entrants to engineering courses have dropped by 14.5% over the period 1985–1998'. Recent trends show no indication that the erosion is not continuing. Hardly a week goes by without the director or chairman of some major company complaining, in the press or on television, that the lack of well trained engineering graduates is impeding growth or new developments. In an article in the *Sunday Herald* (UK) of the 18th May 2008, entitled 'Fuel bosses battle over new recruits', the following observation is made, which crystallises, I think, the looming difficulties for determined expansion of the electricity supply industry:

'With old and new energy sectors struggling in the face of a shortage of graduate engineers and other skilled workers, Bob Keiller, joint chairman of industry association Oil and Gas UK, accused the renewables sector of over playing its importance at the expense of his industry'. Personally given the threat we face, I find it hard to see how the 'renewables sector' could possibly 'over play its importance'.

Even allowing for the laudable enthusiasm for engineering training that exists in the BRIC nations, the global provision of adequately educated and experienced engineering manpower over the next 20 years, particularly those with electrical power engineering expertise, is still liable to fall far short of the numbers required to make possible a massive adoption of renewable technology, as dictated by the requirement both to meet effective emission reduction targets, and to make a rapid transition away from reliance on dwindling fossil fuels. This has to be done sooner rather than later.

Chapter 2
Energy Conversion and Power Transmission

Not believing in force is the same as not believing in gravity.

Leon Trotsky

A raised weight can produce work, but in doing so it must necessarily sink from its height, and, when it has fallen as deep as it can fall, its gravity remains as before, but it can no longer do work.

Hermann von Helmholtz

Ampere was the Newton of Electricity.

James Clerk Maxwell

2.1 Energy Conservation

In order to illustrate the general public's abysmal ignorance of science, the presenter on a quite recent television programme performed a rather simple experiment. As it happens, the experiment illustrates succinctly, what we mean by energy, and how energy relates to power, concepts which are often quite poorly understood. We will need to be clear on these concepts when we begin to examine, later in the book, power budgets for major new energy producing schemes. The presenter was Professor Richard Dawkins, the author of *The Selfish Gene* [1] and many other excellent books on science topics, and the experiment involved a heavy pendulum suspended from the roof of a high lecture theatre.

In the preamble to the experiment Dawkins made it clear to the audience that the suspended metal ball was very heavy, by demonstrating that it took all his time to lift it. Then while maintaining a taught suspension wire he dragged the ball to one side of the room until the ball was at the level of his face and touching his nose. He then let go and stepped aside. The ball of course swung across the room gaining speed as it approached the lowest point of its arc, subsequently rising, slowing to a stop and gaining speed again as it returned to where it started.

The motion was exactly as one would expect for a pendulum. At this point Dawkins stepped smartly forward and caught it. He then asked if anyone in the audience would be prepared to repeat the experiment but without moving away on releasing the ball. Surprisingly there were no takers even when offered a small inducement. Just the merest acquaintance with the first law of thermodynamics, namely the law of conservation of energy, would tell you that that there is no way that the ball would strike you on the way back if you stayed still. Dawkins, of course, demonstrated it himself, not flinching as the ball returned to within an inch of his nose.

You don't have to have lived on this planet for very long to be aware that objects that exhibit weight can possess two types of energy, namely potential energy (energy of position) and kinetic energy (energy of movement). We are ignoring here chemical energy, molecular energy, atomic energy, etc., which manifest themselves only if the heavy object changes its physical form or chemical composition. On drawing back the heavy ball to the height of his nose, Dawkins must do work, which in simple terms is the force exerted against gravity (mass times gravitational acceleration $g = 9.81\ m/s^2$) multiplied by the distance moved. If we neglect frictional effects, this work, in joules, is converted into stored energy or potential energy, also expressed in joules, in the metre-kilogram-second (m.k.s.) system of units [2]. When it is released the ball essentially falls towards the low point of the arc of its suspended swing, losing potential energy while gaining velocity, and hence kinetic energy. Kinetic energy in joules is equal to half the mass of the ball times its velocity squared [2]. At the nadir of its swing all the potential energy supplied by Dawkins has been lost and entirely converted to kinetic energy. In the absence of frictional effects this process would continue for ever if the pendulum continued to do what pendulums do!

This energy exchange between potential and kinetic energy provides a graphic illustration of possibly the most far reaching law in physics, namely the first law of thermodynamics, or the law of conservation of energy. In the absence of any external agency the ball can gain no more potential energy than it started out with and therefore Dawkins had no qualms that the ball would return to his nose but no further. In fact he would know that with some air friction it would stop well short of his nose. Nasal remoulding of pugilistic proportions was not 'on the cards'!

2.2 Power and Entropy

An ideal pendulum, which is not subject to air friction (e.g., pendulum in a vacuum), and which also possesses frictionless hinges (perfect bearings), would oscillate in perpetuity, if allowed to do so, with perfect transference of energy between the potential and kinetic forms. The *total* energy (the sum of the instantaneous potential and kinetic energies) for the ideal isolated pendulum is, however, fixed. No matter how long it is in motion there is no change in the total energy for this closed system formed by the ideal pendulum. The system can be described as 'closed' in a case like this, since it has no influence on the outside world, and the outside world has no influence on it. A bit like a prisoner on Robben Island! Such a system neither delivers nor absorbs power, since power entails an increase or decrease in total energy. Power is defined as the time rate of change of energy [3], and we define an energy change of one joule in one second as a watt in the m.k.s. system of units.

In practice a pendulum system can never be perfect and entirely closed. As the ball travels through the air, friction (collisions between the ball and air molecules) will cause the ball and the surrounding air to warm up. The suspension hinges, if they are not perfect bearings, will also heat up. This heat is an indication that power is being expended by the system. The drag of the air on the ball causes it to lose speed and hence kinetic energy, which in turn means a loss of potential energy. On each swing the pendulum ball will climb less high and eventually the oscillations will cease. A child on a rusty swing will be pretty familiar with the effect. The loss of total energy in the pendulum system can be equated to the heat generated, and power transfer occurs from the pendulum to its surroundings.

The decay in the pendulum motion with time, and the consequential loss of total energy, is a manifestation of the second law of thermodynamics, which simply put states that all systems are subject to increasing disorder or decay and in decaying they lose energy. The technical term that has been coined to encapsulate the process is entropy. Increasing entropy equates to increasing disorder and decay. The original expression of the law, enunciated first by Lord Kelvin, is:

> A transformation whose only final result is to transform into work, heat from a source which is at a single temperature, is impossible.

It really gives expression to a common-sense principle, which, as Steven Weinberg [4] graphically puts it, 'forbids the Pacific Ocean from spontaneously transferring so much heat energy to the Atlantic that the Pacific freezes and the Atlantic boils'.

Few people would bother to ascribe a meaning to the well known nursery rhyme of Humpty Dumpty, but it is really a quite potent, if subliminal, lesson in entropy. The increased disorder of the broken egg that was poor Humpty, could not be restored to order – 'put back together again' even by 'all the king's horses and all the king's men'! If you are an infant or primary school teacher, get your charges to sing it as often as possible, so that one day they may become scientists or engineers! We may desperately need them as fossil fuels vanish.

2.3 Gravity

The pendulum experiment can provide us with one more useful insight into the physics of large scale systems that affect us as humans living on the surface of the Earth. When the heavy pendulum bob is pulled back by Dawkins until it is at the level of his nose we have noted that he must do work against gravity (the attractive force of the Earth which prevents us from disappearing into space!) and that this work is converted to the potential energy now possessed by the ball. The question then arises as to where in the pendulum system does the potential energy reside. If you stretch an elastic band, for example, there is no doubt that potential energy is stored in the taught rubber. If released quickly, the band will fly from your hand as the stored energy in the rubber is rapidly converted to kinetic energy.

In the pendulum, the ball is not squeezed or stretched, and the suspension wire is unchanged, so where is the energy stored? To get students to answer this question I used to ask them to consider what happens when a cricket ball is thrown vertically upwards into the sky. Most people would, I suspect, consider the motion of a thrown cricket ball to represent a relatively trivial science problem, but it is surprising how many students entering university with apparently 'good' physics, can get the dynamics wrong. When asked to draw a picture of the trajectory of a ball rising into the air by depicting the ball at various positions, including the forces acting on it, most students will show the ball correctly slowing as it rises, and speeding up as it falls, by changing the spacing between the representations of the ball or by employing some sort of system of velocity arrows. But, for the vast majority, the upward movement of the ball will be accompanied by force arrows pointing upwards, while the downward motion will be accompanied with downward pointing force arrows. At the point where the ball becomes momentarily stationary some will show a small up-arrow balanced by a small down-arrow. Others will represent gravitational force with some added small down-arrows at various points in the trajectory. When asked why they have shown the forces in this way they will say: 'Well it's common sense isn't it?' However, the reality is, that once released the ball experiences only the downward force of gravity, which is apparently not 'common sense'!

The thrower imparts kinetic energy to the ball in giving the ball an initial upward velocity. If we ignore air friction, which will, as with the pendulum, be relatively insignificant, the ball will slow down as it rises due to downward gravitational force, and as it loses velocity it gains potential energy. The total energy, in much the same way as for the ideal pendulum, will not change as the ball rises. At the highest point in its travel, the ball will be momentarily motionless, all of its kinetic energy converted to potential or stored energy. There will still be a downward gravitational force. In this frictionless case the ball will be materially unchanged during its flight, yet at this position above the Earth it possesses some extra potential energy which it did not have at ground level. Since the ball, and all the molecules of which it is composed, are no different from their ground level state, the added potential energy cannot be stored in the ball itself, so what has changed that could provide the energy storage mechanism? The answer is gravity – there is now a new gravitational field (relative to its ground level value) between the ball and the ground, representing the force of attraction between the Earth and the ball. The potential energy of the ball is stored in this field. In falling back to ground the ball will lose this potential energy as it accelerates under the downward force of gravity, gaining kinetic energy in the process. The gravitational field will return to its original ground value when the ball is retrieved by the thrower. If the Earth's gravitational constant is known, then the constant downward force of attraction between the ball and the Earth is easily calculated by multiplying the mass of the ball with the gravitational acceleration.

These fundamental energy and power relationships, as we shall see, will be very helpful in developing a proper understanding of the essence of electrical power systems in the next section.

2.4 Electricity

The electrical systems (generators, transformers and power lines) which we will encounter in this book are constructed from two types of material, namely metals (conductors) and dielectrics (insulators). To understand what follows it is sufficient to know that in metals, the bound atoms (e.g., copper atoms which have a tiny but 'heavy' nucleus comprising 29 positively charged protons, embedded in a 'cloud' of 29 electrons: I am ignoring neutrons, which seldom make themselves known to electrical engineers!), have one or more electrons weakly attached to the fixed nucleus and these can move 'freely' through the material. Moving electrons represent electrical current and hence metals are 'conducting'. On the other hand dielectrics (e.g., glass or silica, which is formed from the stable element silicon with 32 protons and 32 electrons bound tightly to oxygen atoms with 16 protons and 16 electrons) are materials which have no 'free' electrons and are therefore good insulators. In the m.k.s. system a quantity of charge is expressed in coulombs (C). An electron, which is negatively charged, has a charge magnitude of 1.6×10^{-19} C. *It has no mass.*

In much the same way that it is difficult, in every day life, to be unaware of the effects of gravitational forces, natural electrical forces are also all around us but we are much less conscious of them except in certain special situations. When dry hair is groomed using a comb made of nylon it is not an uncommon experience to hear the wail: 'I can't do anything with it'! The hair strands become charged by the frictional interaction with the comb, and since the 'like' charges deposited on the hair repel, this causes the fuzzy hair effect. Many different insulating materials such as nylon, silk, cotton, plastics can be rubbed together and become charged. What happens is that when two different insulators are rubbed together (hair and nylon) electrical charges, basically electrons, are knocked off the surface of one and are transferred to the other. The material gaining electrons becomes negatively charged, while the material that loses electrons becomes positively charged. Controlled experiments confirm that only two types of charge are involved, namely the negative charge of electrons and the positive charge of protons. The repulsion forces that cause the 'bad hair day' may seem very weak, but in fact a crude comparison with gravity suggests that electrical forces in atoms are vastly larger than gravitational forces by about a billion billion billion billion (one and then 36 zeros) times [5]. So why are we not more aware of these electrical forces if they are so large? Well fortunately materials, whether insulators or conductors, normally have exactly equal numbers of positively charged protons and negatively charged electrons in their molecular structures so that the huge electrical forces of attraction and repulsion between protons and electrons balance out precisely. The numbers we are talking about here are huge because the number of atoms, in a cubic millimetre (about the size of a pin head) of a material such as a metal, is vast – typically about a hundred billion billion. But so perfect is the balance that when you stand near another person you feel no force at all, that can be attributed to electrical charge! If there were the slightest imbalance you would certainly know it. The force of attraction between two people if one of them had 1% more electrons than protons while the other had 1% more protons than electrons would produce a force so great that it would be enough to lift a 'weight' equal to that of the whole Earth!

It is not difficult to find an every day example of natural charge separation that develops very large forces indeed – namely lightning. The process of charge separation in a thunderstorm cloud is rather complicated [5], but in essence it requires a large volume of rising hot air interacting with falling droplets of water or ice particles. The process causes the droplets to become negatively charged by stripping electrons from the warm air, while positively charged air ions rise to the top of the cloud. If the charge separation is sufficient to create a force of attraction between the positive and negative charge layers (within the cloud, or between the cloud and the ground), which is larger than the breakdown strength of air, a violent spark will ensue as the air molecules are pulled apart releasing billions of photons – hence the 'lightning flash'. The energy in a typical discharge is of the order of one thousand million joules! Since the lightning bolts last only a few seconds, power levels of the order of several hundred million watts are dissipated – a watt being a joule/second. This is enough power to boil the water in several thousand kettles all at once! So clearly, very large amounts of energy and power can be extracted from electrical charge separation if only we can control it.

There are basically three ways in which electrical power can be generated controllably. First, there are solar cells (semi-conducting devices), which directly convert electromagnetic waves, usually light, into a constant voltage signal. A large array of solar panels, in which each panel is fabricated from large numbers of semi-conducting junctions, can convert solar energy into usable amounts of direct current and hence electric power. In electrical parlance this is AC/DC conversion where AC is shorthand for alternating current (light is waves and is viewed as AC) and DC equates to direct current. Some small low power electrical devices already employ the technology, such as watches and calculators. The conversion of light into DC current in a semi-conducting junction is, at present, a very inefficient process. It is examined in more detail in Chap. 3 in relation to creating significant levels of power from sunlight. Electrical power can also be generated by chemical processes by means of batteries. For very large levels of power delivery, batteries remain problematic. A fuller assessment of energy storage technology will be broached in Chap. 4.

By far the most effective way of generating large amounts of electrical power is by means of mechanically driven electrical generators. Electrical generators accomplish charge separation, and thereby energy and power, in a controlled and efficient manner, and it is pertinent, for the purposes of this book, to examine how this is done, without delving too deeply into the theory and practice of electrical machines [6, 7]. We will aim to keep it simple, essentially by alluding back to gravity and the pendulum. Energy and power in electrical systems are, rather conveniently, quite similar in form to the corresponding quantities in gravitational systems. It is only a small step from understanding the nature of energy and power in relation to bodies moving under the influence of gravity, to an appreciation of their electrical counterparts. The similarity between the two systems revolves around the fact that while mass and charge are very different, their actions at a distance are not. This similarity then allows us to compare, for each system, how energy is stored and how power is transmitted or delivered.

In the gravitational system, as we have seen, potential energy is created by lifting a heavy mass against the downward force of the Earth's gravity. In a system containing fixed charges (electrostatics) a sphere of charge in vacuum, whether positive or negative, exhibits a force not unlike gravity (electrostatic field). It obeys the inverse square law like gravity, and its strength is proportional to the charge rather than mass as in the gravitational system [8, 9]. If a charged particle of the opposite sign is moved away from the sphere in a radial direction, the force of attraction (electric field) has to be overcome and work is done on the particle – it gains potential energy – just as the cricket ball gains potential energy as it moves away from the Earth. For the charged particle, this energy is stored in the electric field which is created when charges are separated. Once released, the particle will 'fall' back towards the oppositely charged sphere, losing potential energy as the electric field collapses, while gaining energy associated with its motion. This behaviour is not unlike the cricket ball in the Earth's gravitational field. In electrical engineering by the way, potential is essentially synonymous with voltage, which is defined as the work done per unit charge. One volt is the potential energy associated with moving a charge of one coulomb a distance of one metre against an electric field of strength one volt/metre.

In a gravitational system we have already seen, from examination of the motion of a pendulum, that power is released by a mass under the influence of gravity only when it is in motion. The electrical system is no different. Charge, whether positive or negative, has to be in motion before power can be delivered to, or extracted from, the system. Charge in motion implies that a current exists, since electrical current is defined as the rate at which charge is moved – usually inside conducting wires. The unit of current is the ampere and one ampere is defined as one coulomb/second. Since an electron has no mass the energy of the moving negatively charged particle as it 'falls' towards the oppositely charged sphere cannot be kinetic energy which requires a moving mass. So what kind of energy is it? The answer was discovered by Oersted in 1820, but Faraday, Ampere and several other luminaries of the science of electrical engineering have been involved in resolving its nature. In classical electromagnetism, the energy of charge in motion, that is current, is stored in a magnetic field. The relationship can be summarised as: whenever a current flows, however created, a magnetic field is formed, and this magnetic field provides the energy storage mechanism of the charge in motion. The magnetic field basically loops around the path of the moving electron, whatever form the path takes [10]. Magnetic stored energy can be compared to the kinetic energy stored by a moving mass in a gravitational system. Consequently, the pendulum, which oscillates through the mechanism of energy transference from potential energy to kinetic energy and back, can be replicated in electrical engineering by a circuit (an interconnection of electrical components), which permits the transference of energy between that stored in an electric field (electric potential) and that stored when a current and thereby a magnetic field is formed (magnetic energy).

This 'electrical pendulum' is termed a resonant circuit and is formed when a capacitor, which stores electrical energy, is connected to a coil or inductor, which stores magnetic energy. Like the mechanical pendulum, which oscillates at a fixed rate or frequency – about one cycle per second (1 Hz, hertz) in the case of a grandfather clock – a resonant electrical circuit will oscillate at a frequency in the range one thousand cycles per second (1 kHz – in radio terms, vlf) to one thousand million cycles per second (1 GHz – in the uhf television band). The oscillating frequency is dependent on the capacitor magnitude (equivalent to modifying the bob weight in a pendulum) and the inductor magnitude (equivalent to adjusting the length of the suspension wire). In the absence of resistive loss such a circuit would 'ring' forever once set going.

The electrical resonator is an ubiquitous component in electronic systems. It is used wherever there is a need to separate signals of different frequencies. The 'ether' that envelops us is 'awash' with man-made radio waves from very low frequency (vlf) long range signals to ultra high frequency (uhf) television signals to mobile communication signals at microwave frequencies. All receivers, which are designed to 'lock on' to radio waves in a certain band of frequencies, must have a tunable resonant circuit (sometimes termed a tunable filter) at the terminals of the receiving aerial or antenna. Before the advent of digital radios, the action of tuning a radio to a favourite radio station literally involved turning a knob that was directly attached to a set of rotating metal fins in an air spaced capacitor, with the capacitor forming part of a tunable resonator. Thus rotating the tuning knob had the direct effect of modifying the electrical storage capacity of the capacitor and hence the frequency of resonance as outlined above. A dial attached to the knob provided a visual display of the frequency (or the radio wavelength) to which the radio was tuned. Some readers of a nostalgic bent may still possess such a quaint device. In modern digital radios with an in-built processor and programmable capabilities the 'search' function sets a program in operation which automatically performs the tuning.

The above discussion of resonant circuits and tuning may seem a diversion in relation to an explanation of electrical generators, but it has allowed us to, hopefully, move smoothly from energy relations in pendulums, which are almost self explanatory, to the equivalent electrical set up. In the pendulum, its weight has to be moving, and possess kinetic energy, if power is to be transferred to the surrounding medium. In the electrical resonator moving charge equates to current in a coil or inductor, which stores energy in its magnetic field. If the coil happened to be formed from a wire that was not perfectly conducting, heat would be generated in it. This is not unlike, but at a much lower level, the process by which the 'bar' of an electric fire glows hot if sufficient current is forced through it. There is power transference in watts from the energy stored in the magnetic field of the coil to the heat build up in the wire. In this case the resonant circuit would very quickly cease 'ringing' without a continuous stimulus. The analogy with the damped pendulum is not inappropriate here. This comparison is, I have found, singularly helpful to students searching for a robust understanding of the energy/power interplay in an electric circuit.

The energy transfer, or power generation, described above, is from the electrical circuit to the outside world, in the form of heat. In a generator of electrical energy or power we require to convert readily available energy in the form of carbon based fuel, nuclear energy or renewable energy, into electrical power. The key to the conversion process is the magnetic field and the fact that it is formed when charge is in motion. Furthermore, it is helpful to recall a phenomenon that most people will have been made aware of at some stage in their education; namely that when 'like' poles of two bar magnets are brought into close proximity, a force of repulsion is experienced. The magnetic field of a permanent magnet is also associated with moving charge, but in contrast to free electron flow on a wire, here the charge movement is associated with the spin of electrons within the iron atoms. In most materials electron spins are so arbitrarily directed that any magnetic effect associated with this type of charge motion is too insignificant to be meaningful. However, in iron based materials in particular, and also in some other materials, the electron spins can be made to 'line up' (a bit like ballet dancers pirouetting in unison), so that the individual magnetic effects become additive, and a magnet results. The force of repulsion that is experienced when a north pole of one bar magnet is moved towards a north pole of another (or south–south) is caused by a force termed the Lorentz force which arises when *moving* charge is immersed in a magnetic field, or when static charge is immersed in a *moving or changing* magnetic field, although this is more commonly termed the Faraday effect. When like poles are brought close together the magnetic field from one pole produces a force on the spinning electrons within the other pole, which is in a direction tending to drive them out of alignment. In iron, spinning electrons once aligned, are very reluctant to lose their alignment and a secondary force is experienced (the force of repulsion), which is in the direction of preventing further reduction in the distance between the poles.

The Lorentz force is the physical phenomenon, which has made possible the evolution of motors and generators in electrical science, and it can be readily explained by considering the behaviour of a straight conductor when immersed in a magnetic field. Such immersion is usually done, for example, by placing a wire in the gap of a C-shaped permanent magnet. This shape of magnet permits the north pole to be very close to the south pole, and consequently at the narrow gap at the tips of the C, a strong magnetic field occurs. C-shaped or ring magnets vary hugely in size but are generally employed where a steady uniform magnetic field is required. If the aforementioned straight wire is now held at right angles to the magnetic field in the C-magnet gap, and a current is passed through it, the Lorentz force on the moving charge results in a force on the wire, which is in a direction normal (at right angles) to the wire and normal to the magnetic field. This is the 'motor' effect. If the magnetic field strength (or more precisely the magnetic flux density in tesla), the current in amperes and the wire length (metres) are known the force on the wire in newtons is given by the product of magnetic field times current times length [11]. A current of one ampere in a one metre wire, immersed in a magnetic field of one tesla, will produce a force of one newton, which is enough to lift a quarter pound bag of sugar – for readers more used to the imperial system of units! If you prefer the m.k.s. system then this is about 0.5 kg.

If the current carrying wire is disconnected from the external circuit there will obviously be no Lorentz force since there can be no current in an isolated wire, and hence no charge can be moving in the steady magnetic field. Charge movements associated with orbiting or spinning electrons are in entirely random directions in a conductor such as copper, and therefore for these random motions no additive process results and so no force is discernable on the wire. However the conductor, although isolated, still abounds with 'free' charge (about 10^{20} electrons/mm^3) and this free charge (electrons) can be moved through the magnetic field if the wire as a whole is moved. For a wire lying at right angles to the field which is moving in a direction normal to its length, a Lorentz force acts on the free electrons. Almost at the instant that the wire starts moving, free electrons shift to one end of the wire, leaving positive charge at the other. Charge separation occurs, which ceases as soon as the resultant electrostatic force (between the positive and negative charge clusters) just balances the Lorentz force. This happens in a very small fraction of a second. The resultant charge separation means that a voltage exists between the ends of the wire *while it is in motion* in the magnetic field. It disappears as soon as the wire stops moving. This induced voltage is commonly referred to as the electromotive force (emf) in the moving wire, since it derives from the Lorentz force [11]. What we now have is generator action in its simplest form. For a one metre long wire moving at right angles through a one tesla magnetic field at a velocity of one metre/second, an emf of one volt is generated. In practical generators much higher voltages are possible by series connecting together multiple moving wires. The simplest way of doing this is by forming a winding on an armature and rotating it at high speed so that the 'wires' forming the winding cut through the strong fluxes of static ring magnets, with multiple poles wrapped around the armature, as is done in DC generators, or alternatively by rotating armature mounted magnets so that their fields sweep over fixed stator windings as in AC synchronous or induction machines. The two processes are essentially equivalent. We shall talk about synchronous and induction machines later in relation to electric power generation.

2.5 Generators

Life today is almost unimaginable without mains electricity. It provides lighting for houses, buildings, streets, supplies power for domestic and industrial heating, and for almost all electrical equipment used in homes, offices, hospitals, schools and factories. Improving access to electricity worldwide has been a key factor in 'oiling the wheels' of modern life. With the brief insight into the relevant physics and engineering that was furnished in previous sections we are now in the position to take a quantitative and critical look at electricity generation as it is currently practised around the world.

In modern oil, gas, coal, hydro-electric or nuclear power stations, the generator set is usually of the synchronous type (Fig. 2.1). This means, that in the context of rotating electrical machines, it is of the 'inverted' type of construction, as mentioned in the previous section, where the windings supplying the electrical power are stationary, being wound onto the stator, while the armature (the moving part) houses the rotating magnetic stack. A key feature of this set up is that the generated voltage is alternating (AC) and its frequency is strictly controlled by the rotational speed of the machine. The frequency of the supplied AC output of a given machine is not difficult to estimate. It is given by the product of the number of rotor poles (magnetic poles) and the rotational speed in revolutions/minute, divided by 120. A four pole machine (probably the most common arrangement) rotating at 1500 rpm will generate a 50 Hz supply voltage. In the USA where the electricity supply is set at 60 Hz, the speed of the machine has to be 20% higher. Maintaining the frequency of the AC output within acceptable limits means that the prime mover (engine or turbine) must be governed to hold its speed constant to within 3–4% of the optimum value. The generated voltage for a single phase machine is given by a winding constant (~4.15) multiplied by the number of turns

Fig. 2.1 Modern steam turbine driven generator system

(i.e. in the stator winding) multiplied by the flux under each magnetic pole multiplied by the frequency [11]. A single phase machine is one which essentially has only one stator winding with two output terminals. The supplied voltage is a single alternating signal at the design frequency of the machine. More power with higher efficiency is delivered if the stator carries more than one winding; the norm is three, in which case there are three live output terminals, plus a neutral connection, delivering three phase power. This means that between any two terminals there is a 50 Hz sinusoidal signal as in a single phase machine, but the phase voltages, as they are termed, while of similar magnitude, are shifted in phase relative to each other by ± 120°. The nature of three phase supply and its application is not significant in relation to the environmental impact of electrical systems, once system losses have been established, and we will not need to refer to this generation mode again, other than to provide some background for discussions on electricity transmission.

Output voltages of between 20,000 and 30,000 V are typical of synchronous machines installed in modern fossil fuel power stations. The fundamental requirement of large power station generators to deliver about 30,000 V (30 kV) at 50 Hz (or 60 Hz) exercises a considerable constraint on the design of a synchronous machine and in visual terms structural differences between machines can seem marginal. The primary variable is power capacity, and since power is essentially volts multiplied by amperes, it means more powerful machines have to be capable of sustaining higher currents when on full load. High currents incur heat, which means bulkier windings, and more effective cooling is required to minimise energy loss. Inevitably the machines become bigger, although structural constraints of a mechanical complexion are enforcing a halt to the trend. Fossil fuel power stations are capable of generating power levels to the grid anywhere between 100 and 2000 MW. If we take an intermediate figure of 500 MW, a synchronous machine capable of supplying this sort of power will be of the order of 30 ft long, and 12 ft in diameter. Such a machine would typically supply 21,000 A at 24,000 V, at a frequency of 50 Hz (60 Hz) as it spins at between 3000 and 3600 rpm.

Unfortunately, not all of the power provided by the prime mover is converted into electrical power to the grid in a synchronous generator. There are a number of sources of power loss which cannot be circumvented, although some effort is made to keep them as low as possible. These unavoidable loss mechanisms are conductor losses, core losses, mechanical losses and stray losses.

Conductor loss, or ohmic loss, occurs whenever current is forced through a wire. It comes about because real wires formed from copper or aluminium, for example, exhibit some resistance (ohms) to free electron flow (current in amperes) through them. Electrons flowing along a wire can very crudely be likened to balls rolling down a pin-ball machine. In the pin-ball machine the balls, as they travel, collide against the pins, which divert them from the direct path to the bottom of the table. After many collisions the total distance travelled by a typical ball will be much more than the length of the table. On colliding with a pin, a ball will actually lose a tiny amount of kinetic energy, which will appear as vibrational energy in the pin. The movement of an electron through a wire is not unlike this. Fixed copper or aluminium atoms (the pins in the pin-ball machine) present obstacles to the electrons flowing through the wire. The distance travelled by the average electron is generally much longer than the direct path through the wire, and the collisions between the electrons and the fixed atoms generates atomic scale vibrations. This atomic agitation manifests itself as heat. Only at absolute zero temperature (0 K) are the atoms of a material completely still. In a generator, if the winding current is known and if the resistance of the winding has been measured the heat loss in watts is given by amperes-squared times winding resistance [12] multiplied by the number of windings. For a typical synchronous generator this copper loss is approximately 3% of the power delivered. In a 500 MW machine, 15 MW is dissipated in the windings. This is enough to boil the water in 15,000 brim full kettles, or to power, across Europe, two high speed electric trains!

The windings in any electrical machine are usually wrapped around cores made of soft iron. 'Soft iron' is a form of iron that is easily magnetised or demagnetised (spinning electrons are easily aligned or misaligned) and it is used to maximise magnetic flux through the windings of the machine. These cores form the armature and stator of the machine. In an operating generator the large currents flowing in the windings induce magnetic fields in the cores. These alternating magnetic fields, through the Faraday effect, which is actually quite similar to the generator effect, induce secondary currents within the core structure of the generator. These currents are commonly termed eddy currents. Since iron is not a perfect conductor, resistive losses associated with the eddy currents also occur within the metal (iron) structure of the generator. Most people who have ever used a battery charger will have been aware that the charger gets warm. This is because the charger contains a transformer, which has an iron core encased in multi-turn windings made from insulated copper wire. The heat has the same source as in the generator – namely copper loss and core loss. If you are very perceptive you may also have observed a faint hum coming from the charger. The laminated core of the transformer, which is laminated to minimise core loss, can vibrate (hence the hum) because of relative movement between laminations. The movement is driven by the Lorentz force described earlier (essentially the motor effect). This process absorbs energy and adds to power loss. It also occurs in AC power generators and is considered to be part of the core loss of the machine. This loss is generally in the vicinity of 4% of the generator output – enough to heat 12,000 platefuls of Scots' porridge in 12,000 microwave ovens! Even if you like porridge: not a great idea.

Mechanical losses mainly include drag effects due to air compression and air friction, which occurs in the air gaps between the rotor and the stator when the rotor is revolving at high speed. Bearing losses also come into this category. In total, power dissipation of a mechanical nature can contribute a further 4% reduction in machine efficiency. Stray losses describe all the other miscellaneous losses that do not fall into the above 'pigeon holes', and although small they represent a finite addition to inefficiency. These losses are generally estimated to contribute about 1% to the total. A generator with a 500 MW rated output power will, because of these losses (12% of 500 MW equals 60 MW), require a turbine or diesel engine delivering 560 MW of mechanical power at its input shaft. The generator efficiency is then (500/560) × 100 = 89.3%. For most power station generators the assumption of a figure of 90% for generator efficiency would not be far off the mark for the purposes of assessing environmental impact.

We also need to give some attention to the 'prime mover' in any generation station. Today about 86% of the world's electricity is generated using steam turbines fuelled mainly from coal and oil. Most of the rest is provided by nuclear power, which also generates electricity through the agency of steam. We can therefore limit our attention to this type of prime mover when we later use conventional power generation as a benchmark for assessing generation from renewable sources. The steam turbine is designed to extract thermal energy from pressurised steam, and convert it into useful mechanical power output. It has almost completely replaced the long-lived reciprocating piston steam engine, primarily because of its greater thermal efficiency and higher power-to-weight ratio. Also, because the turbine generates rotary motion directly, rather than requiring a linkage mechanism to convert reciprocating to rotary motion, it is particularly suited to the role of driving an electrical generator.

In thermodynamics, the thermal efficiency (η_{th}) is a dimensionless performance measure of a thermal device such as an internal combustion engine, a boiler, or a furnace, for example. The input to the device is heat, or the heat-content of a fuel that is consumed. The desired output is mechanical work, or heat, or possibly both. Because the input heat normally has a real financial cost, design engineers close to the commercial realities tend to define thermal efficiency as [13] 'what you get' divided by 'what you pay for'.

When transforming thermal energy into mechanical power, the thermal efficiency of a heat engine is obviously important in that it defines the proportion of heat energy that is transformed into power. More precisely, it is defined as power output divided by heat input usually expressed as a percentage. The second law of thermodynamics puts a fundamental limit on the thermal efficiency of heat engines, such that, surprisingly, even an ideal, frictionless engine cannot convert anywhere near 100% of its input heat into useful work. The limiting factors are the temperature at which the heat enters the engine, T_H, and the temperature of the environment into which the engine exhausts its waste heat, T_C, measured in kelvin, the absolute temperature scale. For any engine working between these two temperatures [13] Carnot's theorem states that the thermal efficiency [14] is equal to or less than one minus the ratio T_C/T_H. In essence, what this means is that we cannot extract more heat from the steam or fuel than is permitted by the dictates of the second law and the requirements of entropy. For T_C to be lower than the ambient temperature we would be requiring a lowering of entropy. Ice, for example, has less entropy than warm water vapour at room temperature. However, it requires power input to form ice as we know from the cost of refrigeration, unless we live in Antarctica! The limiting value is called the Carnot cycle efficiency because it is the efficiency of an ideal, lossless (reversible) engine cycle, termed the Carnot cycle. The relationship essentially states that no heat engine, regardless of its construction, can exceed this efficiency.

Examples of T_H are the temperature of hot steam entering the turbine of a steam power plant, or the temperature at which the fuel burns in an internal combustion engine. T_C is usually the ambient temperature where the engine is located, or the temperature of a lake or river that waste heat is discharged into. For example, if an automobile engine burns gasoline at a temperature of $T_H = 1500°F = 1089\,K$ and the ambient temperature is $T_C = 70°F = 294\,K$, then its maximum possible efficiency [14] is $\eta_{th} = \left[1 - \frac{294}{1089}\right] \times 100 = 73\%$.

Real automobile engines are much less efficient than this at only around 25%. Combined cycle power stations have efficiencies that are considerably higher but will still fall at least 15 percentage points short of the Carnot value. A large coal-fuelled electrical generating plant turbine peaks at about 36%, whereas in a combined cycle plant thermal efficiencies are approaching 60%.

In converting thermal energy into electrical power, we can therefore conclude that in a conventional modern power plant in which a steam turbine drives a synchronous generator, the conversion efficiency will be at best $0.6 \times 0.9 \times 100 = 54\%$ in a power station operating in combined cycle mode where waste heat is used to warm the houses of the local community. A coal fired power station that delivers 500 MW of electrical power to the grid dissipates almost the same amount in the form of heat. If you have given some thought to thermodynamics and the second law, you will not be surprised to learn that this just leaks inexorably into the atmosphere!

2.6 The Grid

The final element in the electricity supply 'jig-saw' is transmission and distribution. In the electricity supply industry transmission and distribution are viewed as quite separate activities. In the industry, when they talk of transmission, the bulk transfer of electrical power from several power stations to towns and cities is being considered. Typically, power transmission is between one or more power plants and several substations near populated areas. The transmission system allows distant energy sources (such as hydroelectric power plants) to be connected to consumers in population centres, thus allowing the exploitation of low-grade fuel resources that would otherwise be too costly to transport to generating facilities. Electricity distribution, on the other hand, describes the delivery of electricity from the substation to the consumers.

A power transmission system is sometimes referred to colloquially as a 'grid'; however, for reasons of flexibility and economy, the network is not a rigid grid in the mathematical sense. Redundant paths and lines are provided so that power can be routed from any power plant to any load centre, through a variety of routes, based on the economics of the transmission path and the cost of power. Much analysis is done by transmission companies to determine the maximum reliable capacity of each line, which, due to system stability considerations, may be less than the physical or thermal limit of the line.

Owing to the large amount of power involved, transmission at the level of the grid normally takes place at high voltage (275 kV or above in the UK). This means that a transformer park exists at all power stations to raise the generated voltage, which is typically at about 25–30 kV, up to the local grid level, i.e., about ten-fold. This process adds, through ohmic losses in the transformer windings and core losses in its magnetic stack, possibly a further 1% to generation losses. However, transformation to high voltage is essential to avoid much higher losses in the cables of the grid system. This is easily explained by recalling that the magnitude of conductor loss [12] in a wire is proportional to the resistance of the wire and to the square of the current through it. If the grid is required to carry a power (P watts), which is largely equal to the transmission voltage (V) multiplied by the transmission current (I), then by increasing V ten-fold the current can be reduced by a factor of ten for the same power and hence the cable losses by a factor of one hundred. This is a very considerable saving on wires which could be hundreds of miles long.

In calculating the power loss in very long electrical cables it is easy for the unwary engineer to under-estimate its magnitude, because of a phenomenon called 'skin effect'. It is, therefore, sensible to take a short detour here to explain this phenomenon because it is important to some of the transmission issues that will be addressed later. Suppliers of electrical materials are required to perform rigorous testing programmes on their products, to provide users with accurate values for the physical constants of the materials purchased. Examples of these constants are thermal conductivity, specific heat, electrical conductivity, electrical resistivity, permittivity, permeability, etc. The tendency of a material to resist the flow of electron current is represented by its electrical resistivity (ohm-m), or its reciprocal, electrical conductivity (siemen/m). The resistance of a conducting wire in ohms (Ω) is then given by resistivity times length divided by cross-sectional area [12], provided the current flow is unvarying (DC). For example a DC cable, comprising two 156 mile long lengths of 5 cm diameter hard aluminium wire, for which the resistivity is 2.86×10^{-8} ohm-m, has a resistance of almost 1.2 Ω. This is a very low resistance. Even so, a DC current of 100 A in this cable would generate 12 kW of loss in the form of heat.

If the cable carries a 50 Hz AC signal the above calculation would be erroneous because of the troubling (to students) quantity termed skin effect. So what is skin effect and how do we adjust the calculation to accommodate it? In Sect. 2.4 you will remember that we discovered that electrical energy is stored in electric and magnetic fields. Since power is rate of change of energy, it follows that when we transmit power (move energy) through transmission lines or across space (radio waves) the agency that allows us to do this must be electric and magnetic fields. The transport mechanism takes the form of electromagnetic waves. When AC power is transported through a transmission line, the power is not carried through the interior of the wires, but in the space between the wires as an electromagnetic wave. If transmission system wires could be made perfectly conducting, it would in principle be possible to carry high electrical power along filamentary wires with infinitesimally small cross-sectional area. If their cross-section is tending towards zero in engineering terms (i.e., infinitesimally small but not sub-atomic dimensions), it can be concluded that the finite power being transmitted across the grid cannot be propagating in the interior of the wires, otherwise we have the physically impossible scenario of power density tending towards infinity! Again, the important point to reiterate here is that the power is carried through the space between the wires on an electromagnetic wave, and the wave cannot penetrate into the perfectly conducting wires. A perfect conductor is, in fact, a 'perfect mirror' for electromagnetic waves. In this case we can state that the 'skin depth' is zero, and that the current in the wires flows in an infinitesimally thin, 'atomic thickness', surface layer. There are still plenty of 'free electrons' within this 'atomic' layer to accommodate the finite current. In non-perfect conductors such as copper, electromagnetic wave penetration into the interior is possible, but the wave attenuates very rapidly. The skin depth (δ m) for real materials is defined as the distance from the surface at which the penetrating wave has diminished to 1/e of its surface magnitude, where the exponential constant $e = 2.718$. For aluminium at 50 Hz the skin depth is 17.2 mm. At this frequency skin depth could raise the resistance of the 100 mile long aluminium cable by 50% above the DC value of 1.2 ohms. In practice the suspended grid wires are designed to have a diameter of little more than the skin depth in order to minimise weight, in which case there will not be much difference between the DC and AC resistances of a power line. The line resistance per phase of high voltage grid is typically 0.7 ohms/mile. Given that the line will carry a current of around 300 A we end up with a 'ball-park' figure for the power transmission loss for the grid of 200 kW/mile or 124 kW/km, a statistic that we shall find useful later. In terms of percentage of the power carried, this represents a power loss of 8% per thousand kilometres for a 750 kV line.

The pylon-supported wires of the grid are almost exclusively formed from aluminium laced with a core of steel strengthening strands. Although aluminium is a poorer conductor than, for example, copper it is preferred in this role because it has a much better conductivity to weight ratio making it lighter to support. The steel core, which gives the aluminium wire enough strength to be suspended over long distances, has no electrical effect because of skin depth. There is a limit to how high the voltage can be raised to diminish conductor loss and this is set by air breakdown or corona discharge, particularly in humid or wet conditions. If corona occurs, losses can escalate markedly.

Underground power transmission over long distances is not really an option, unless exceptional circumstances exist, due to its high cost of installation and maintenance (about ten times more costly than overhead cables). It is really only used over short distances, normally in densely populated areas. There is also a significant technical problem in the transmission of AC power with underground or undersea cables. Since they are usually of coaxial construction, this means that they are subject to what the industry describes as high reactive power loss. Very long over ground transmission lines, suffer the same phenomenon. It sounds complicated but all it really means is that in buried cables the current carrying conductors (three for three phase transmission) are embedded in an insulating material, which is essential in order to maintain the more closely spaced conductors at a constant separation. This construction results in much higher capacitance per unit length than occurs with pylon supported transmission lines. Capacitance tends to increase with the surface area of the current carrying conductors and to decrease with separation distance, and high capacitance as we have already seen, equates with high electrostatic energy storage, which in turn implies high charge accumulations. The reactive power phenomenon associated with the high capacitance, or high storage capability, of long transmission lines is quite difficult to explain by a gravitational analogy because we cannot include phase effects. Nevertheless we can illustrate the nature of the difficulties that are introduced by the presence of a storage element in a transmission system.

Through the agency of gravity, as we have seen, water in an elevated reservoir (generator) if released into a descending trough or channel (transmission system) towards a water wheel (load/consumer) at a lower level, will transmit power to the wheel. Some energy in transmission will be lost through frictional losses in the channel and inefficient power collection by the water wheel, but the process is relatively uncomplicated. Let us consider now what happens if a second smaller reservoir exists between the original reservoir and the water wheel. If the low reservoir is initially empty, water will flow down the channel simply transferring potential energy from the upper store to the lower store with no power going to the water wheel. Nevertheless, despite the fact that no power reaches the turbine, frictional power loss in the channel continues to occur due to the water rushing to fill the lower water store. It is only once the lower reservoir is full that power gets to the wheel. In electrical terms, with very long transmission lines exhibiting high capacitance (storage capacity), power from the generator is initially used simply to store energy in the grid, resulting in high power loss as current surges into the system. Once energy transfer to the line capacitance is complete (in milliseconds) power will flow steadily to the consumer. This is DC in action. On an AC line the situation is much worse. In terms of the water channel analogy we now have to empty the lower reservoir quickly and completely through an alternative outlet on a regular basis, to get an idea of the problem with long lines and AC transmission. On a very long line the charging current can reach a magnitude that produces excessive ohmic loss in the wires, and the conductors may be raised in temperature to beyond their thermal limit. This means that AC cannot be transmitted across power lines that are more than a few hundred kilometres long, or over long undersea cables, without reactive power compensation. Not surprisingly there is considerable reluctance in the power supply industry to adopt buried or undersea transmission lines, unless there is no alternative. In this case DC becomes the preferred mode of transmission.

The grid itself, as indicated earlier in this section, is a loose interconnection by means of transmission circuits, of multiple power stations and loads (consumers through the agency of substation transformers). This interconnection places stringent frequency and voltage constraints on power suppliers, but this is off-set by versatility of supply, higher operating efficiencies and economies of scale.

The need for frequency and voltage control for grid connected generators can probably best be illustrated by considering a very simple electrical circuit formed from batteries and loads (light bulbs for example). The arrangement is valid insofar as we know that at 50 Hz, wavelength on the grid is so long (~ 6600 km), that branches of the grid (typically 100 km) are sufficiently short in wavelength terms for the voltage and current on the line to be considered to have DC characteristics. 'Power stations' on our elementary 'grid' circuit each have two rechargeable batteries connected through a reversing switch. There are several power stations and several loads (consumers) all interconnected by two copper wire loops, one 'live' and one 'earth'. The power station batteries are connected through a switch such that one (battery A, say) has its positive terminal connected to live with the negative terminal connected to earth, while the other (battery B) is disconnected. When the switch is reversed battery B is connected to the circuit but with the positive and negative terminals swapped over. The bulbs are connected at various positions around the loop with one terminal connected to the live wire and the other to the earth. Provided the reversing switches are synchronised, so that all A batteries are connected to the loop at the same time, power will flow from the batteries to the bulbs, lighting them up, if the batteries are all at the same voltage level. Reversing the switches to connect batteries B to the loop changes nothing electrically provided the switch over is synchronised. In this case the light bulbs will seem to glow uninterrupted. If the reversing switches are all automatically vibrated synchronously at 50 Hz we have essentially what happens on the full scale grid. Clearly it is important that when a new battery pair is to be connected to the circuit, the reversing switch is being vibrated at exactly the same rate as all other switches on the circuit and that it is synchronised so that the positive terminal of battery A is connected to the loop when the loop is positive – and vice versa for battery B. If this does not happen, a high current will flow from the circuit to the battery, possibly causing failure. On the grid, power station failure is unlikely if synchronisation is not achieved, but considerable power loss and instability will result. Our battery circuit also shows us that it is important that all the battery 'stations' supply the same voltage to the loop, otherwise even if synchronised to the loop voltage, a battery pair that is low in voltage will absorb power from the loop – charging up the battery pair. The same is true for the full scale grid, with power flowing from the grid to a power station, if its voltage does not match that of the grid. Finally, in the circuit model if bulbs are removed or added to the system – demand fluctuation – the batteries cope automatically. This is not true of power stations linked to the grid. Demand prediction and control is essential to efficient operation of the system. Synchronisation, voltage control and adjustment to demand, are very important in electrical power systems coupled to the grid, and as we will see, this can present significant difficulties for renewable power sources. For electricity distribution from the grid to cities, towns, factories, hospitals, schools, homes etc., down conversion at substations from 400 to 33 kV, or less, is required. Substation transformer losses add a small but finite contribution to overall transmission and distribution losses.

2.7 The Power Leakage Dilemma

Losses in the grid are estimated to be of the order of 7% in USA and Europe. Consequently, at the point where electricity is distributed to end users, only between 30 and 50% of the original power supplied to the system, in the form of heat from steam in a conventional generating station, is available. The two figures depend on whether an estimate of 36% or 60% is adopted for turbine efficiency. This implies that at best only *half* of the energy in the coal or oil burned in a modern fossil fuel fired power station produces useful electricity for the consumers. The rest goes 'up in smoke'!

The situation is actually worse than this if we consider end user applications, which are rarely efficient. However, before we can address this issue, a method of quantitative estimation with a good scientific pedigree is required. The technique is best illustrated through a tale attributed to Enrico Fermi, one of the giants of twentieth century science, and a leader of the Manhattan Project during World War II. The story is told that at stressful moments during the project he would fortify morale of his fellow bomb makers by throwing out quirky mental challenges [15]. 'Out of the blue' he might ask 'How many piano tuners are there in Chicago?' Fermi took the view that any good physicist, or any good thinker for that matter, should be able to formulate a logic based reasoning procedure for attacking any problem and come up with an answer which is within 'an order of magnitude', that is within a factor of ten, of the correct one.

So what does a reasoning process that gets to an approximation of the answer to the Fermi question look like, when the piano tuning trade is a rather obscure one, and when the challengee's knowledge of Chicago is no more than that it is located on the western side of Lake Michigan? The method is outlined in Angier's book [15] with help from Laurence Krauss [16]. The reasoning process, with its scientific underpinning, goes as follows. Chicago is one of the largest cities in the USA, which means its population must be up in the multi-million range, but not as high as eight million, the population of New York. Let's give it four million. How many households does this amount to? Say four people per dwelling, which means one million households. How many of these one million household are likely to possess a piano. If one considers the rate of piano ownership among ones friends a figure of about 10% would probably be not unreasonable. So we are looking at something like 100,000 pianos in Chicago that will need an occasional tune up. What do we mean by occasional? A tune up about once a year seems a reasonable guess, at a fee of say $75 to $100 per visit. So at this rate how many pianos will a professional piano tuner have to tune per year to stay solvent? Possibly two a day, which amounts to ten a week, for a five day week. In a year he will have to tune 400 to 500 pianos. Hence, if we divide 100,000 by 400 or 500 we get a ballpark estimate of 200 to 250 piano tuners in Chicago. The actual answer is apparently 150. Quantitatively, this is a very good estimate, being well within the 'order of magnitude' criterion.

Returning to the electrical power leakage problem, we can confidently state that the primary end users of electricity are industry (21%), energy supply (6%), commerce (13%), public sector (35%) and homes (25%) [10]. In industry we also have the statistic [17] that almost 66% of the electricity is used to drive motors and supply transformers. As we now know from earlier calculations, the efficiency of electrical devices of this genre is of the order of 90%. Most of the remaining 34% of electricity distributed to industry can reasonably be assumed to be employed in lighting. Again presuming that industry largely uses fluorescent lighting we can estimate that their lighting efficiency (proportion of input power converted to light) is probably of the order of 10%. So we can say that industry wastes $0.1 \times 0.66 + 0.9 \times 0.34 = 0.372$, that is 37% of the electrical power supplied to it.

In commercial offices, and in the public sector (schools, hospitals, offices, etc.), the vast majority of the supplied electrical power will be for lighting and air-conditioning, with a small proportion used to operate computers and appliances. An estimated overall efficiency of the order of 25% is not unrealistic and consequently these sectors dump to the environment as heat, 75% of the power supplied to them. In the domestic sector in the UK, the way in which electricity is allocated, on average, to different activities (post-2000) is available in government literature [17]. 25% is used in lighting – probably mainly incandescent bulbs which are very inefficient at just 2% of the power used appearing as light. 24% is employed in supplying fridge/freezers, which are rather inefficient because of poor insulation although the pump motor and power supply will also make a substantial contribution to a typically quoted figure of 12%. Washing machines, dryers, dishwashers account for 18% of usage. Here losses can be attributed to power supply transformers, pump motors, bearings and expelled hot water giving a typical efficiency of about 60%. Cookers, kettles, microwaves take up another 18%, and while microwave ovens are reasonably efficient at 70%, cookers and kettles are not – so an overall figure here of 50% is probably not too wide of the mark. Finally, consumer electronics, which have penetrated the domestic market in a big way in recent years, now accounts for 15% of electrical power usage in the home. Here power losses are mainly attributable to power supplies, which can deliver efficiency levels as low as 20% and as high as 80%. A mean of 50% is probably not unrepresentative. Of the electric power supplied to homes in the UK the power wasted is as follows:

- lighting $= 0.25 \times 0.98 = 24.5\%$;
- fridge/freezers $= 0.24 \times 0.88 = 21.1\%$;
- washing-machines $= 0.18 \times 0.4 = 7.2\%$;
- cookers $= 0.18 \times 0.5 = 9\%$; and
- electronics $= 0.15 \times 0.5 = 7.5\%$.

This gives an accumulated electrical power loss in the typical home in the UK (the number will probably apply in most industrialised nations), of at least 70%. Space heating is ignored for this exercise since few homes (except perhaps in nuclear France?) are heated using electricity at the present time. The total wastage in all sectors is therefore as follows:

- industry $= (0.21 + 0.06) \times 0.37 = 10\%$;
- commerce $= 0.13 \times 0.75 = 9.7\%$;
- public sector $= 0.35 \times 0.75 = 26\%$; and
- homes $= 0.25 \times 0.7 = 17.5\%$.

This represents a total percentage loss by end users of 63%. Of the power contained in the heated steam entering the turbines of a combined cycle electricity generator only, at best, $(0.5 \times 0.37)100 = 19\%$ results in usefully employed power by the end user. For generators operating in non-combined cycle mode the figure is even lower at 11%, with 89% of the original power being wasted.

Clearly the accumulated power dissipation involved in generating electricity, in transporting and distributing it, and in the way we use it, is almost mind boggling in its magnitude. At present, the saving grace is that electricity represents only 10% of energy usage in industrialised societies. In the future when electrical power provides 100% of our needs, this level of misuse of precious resources is likely to be viewed as quite irresponsible and unacceptable! It seem utterly perverse that governments are planning to move to a world based on renewables, supplying power to consumers through the agency of electricity, without seriously addressing the vast wastage that currently applies in the electricity industry? Not all of the inefficiencies are surmountable, of course, but obviously the more efficient we can make electricity supply and distribution, and the more efficiently we use it, the less pressure will be placed on the drive to exploit renewable resources, and the less will be the environmental damage.

The issues involved in supplying our energy needs from electricity, when these are generated from renewables, are addressed in the next chapter.

Chapter 3
Limits to Renewability

Someday we will harness the rise and fall of the tides and imprison the rays of the sun.

Thomas Edison

The world today is full of problems that cannot be solved by the type of thinking which we employed when we created them.

Albert Einstein

For a successful technology, reality must take precedence over public relations, for Nature cannot be fooled.

Richard P. Feynman

3.1 Power from the Sun

The source of all renewable energy on planet Earth is predominantly the solar system. In pondering this bountiful energy provider, it is humbling to contemplate the elegance of the physical insights that have evolved by dint of human effort over the past 200 years, allowing us now, to fully comprehend it all. Quantum theory enables us to understand the nuclear processes in the sun; electromagnetic theory tells us everything we need to know about radiation from the sun, while the general theory of relativity gives us complete insight into planetary behaviour. The scientific journey of discovery which has revealed the nature of all things physical has been elegantly recorded by many scientists, but Steven Weinburg's [1] book *Dreams of a Final Theory*, has illuminated it most completely and effectively for me. What has yet to be explained, such as the relative weakness of gravity in relation to the electromagnetic and nuclear forces, is not going to impinge on the practice of science and engineering on planet Earth in the foreseeable future. While the fortuitous arrangement of our sun and planets so beautifully, and perhaps precariously, provides mankind with an irreplaceable life support system, it also gives mankind so much to carelessly exploit.

A.J. Sangster, *Energy for a Warming World*,

In Chap. 2 we have already explored aspects of gravitational theory at the Newtonian level, which is more than adequate for explaining earthly phenomena. The gravitational physics we learned there will be employed in Sect. 3.2 of this chapter to quantify the potential capabilities of hydro-electric schemes. Anyone who has visited a natural waterfall such as Niagara Falls can hardly fail to have been impressed by the shear power of the plunging water. On the face of it, this power is endlessly exploitable at no cost to the environment. It is supplied, for free, by gravity. It is made possible by solar warming of the atmosphere generating rain, which in turn fills conveniently elevated lakes and reservoirs. Whether or not man-made reservoirs are 'for free' is another matter as we shall see.

The dynamics of wind can readily be illuminated through science by applying the second law of thermodynamics. It predicts that a warm fluid will always drift towards a cooler place, thereby increasing its entropy. Most people will have observed the effect when they warm a pan of water. The warmer water over the heat source induces a convection current as it mingles with cooler water. In atmospheric terms, the fluid is air, and the troposphere is differentially warmed by the sun as the Earth rotates and wobbles on its axis. The resultant convection currents form areas of high and low pressure within the atmosphere and the pressure gradients provide the mechanism for planetary winds, which can range in speed, as we well know, from a few miles per hour (mph) to well over 100 mph. The sun, through the agency of the wind, is also the source of the energy contained in ocean and sea waves. These waves are mechanically sustained surface waves that propagate along the interface between water and air. The restoring force that underpins the wave dynamics is, once more, provided by gravity, and so ocean and sea waves are often referred to as surface gravity waves. As the wind blows, the equilibrium of the ocean surface is perturbed by wind generated pressure and friction forces. These forces transfer power from the air to the water, which is transported by the water waves. Extraction of electrical power from the wind and the waves will be examined in Sects. 3.3 and 3.4.

While tidal (Sect. 3.5), solar (Sect. 3.6) and geothermal (Sect. 3.7) phenomena perhaps represent more predictable sources of power, cost effective exploitation in the kind of quantities that energy hungry human societies will demand, will as we shall see, present a serious barrier to their widespread use. Tidal power is, of course, provided by the gravitational forces of the moon and the sun acting on the volume of water contained in the Earth's oceans and seas. The moon pulls at the ocean/sea surface displacing it fractionally against the much stronger earthly gravitational force. At full-moon and new-moon the gravitational tug is largest, since the moon and sun are aligned, thus producing the most significant ocean displacements (spring tides). Tidal variations of as much as 50 feet occur in some parts of Canada. It is, however, the shear volume and mass of the water that is being raised over vast areas of ocean by these tidal movements that provides us with such large amounts of potential energy. Some of this energy is eminently exploitable. But again we need to ask – how much?

To understand solar power and how it may be possible to collect it efficiently, it is necessary to have some knowledge of electromagnetic waves. Nuclear processes in the sun, as it converts through fusion hydrogen into helium, result in the emission of electromagnetic waves (photons) over a range of frequencies from below infra-red to above ultra-violet, with a lot of light in between. These waves spread out radially from the sun, in all possible directions, but those contained within the notional cone, centred on the sun and subtended by the disc of the Earth, are responsible for delivering solar power to the Earth. Scientists can estimate quite precisely how much radiant power is emitted by the sun and they know that the power density of this radiation in W/m^2 diminishes as the inverse of the square of the radial distance from it. Consequently, employing some not very taxing geometry based calculations it is possible to compute the solar power striking the Earth averaged over its disc (the solar constant). The most recently available figure is $1367\,W/m^2$. Given that the Earth's diameter $D = 12760$ km, and that the disc area $\frac{1}{4}\pi D^2$ is therefore $1.28 \times 10^{14}\,m^2$, it is not difficult to work out [2] that the solar radiant flux on the Earth's atmosphere exceeds 170,000 TW, or 5.7×10^{24} J/year. Surprisingly this is more than estimates of the remaining worldwide fossil and nuclear fuel energy resources combined. Calculations vary, but fossil fuel reserves (gas, oil and coal) are thought to amount to an estimated 0.4×10^{24} J while nuclear fuel such as uranium is potentially capable of yielding 2.5×10^{24} J. Clearly solar energy flux is an abundant renewable resource, but, as we shall see, effective collection in very large quantities is subject to a range of constraints. Perhaps mankind would have discovered how to do it by now, if it hadn't been for more easily exploitable fossil fuels?

Geothermal schemes for electricity generation are predicated on accessing the energy buried in the hot interior of the planet. The planet's internal heat source was originally generated during its formation several billion years ago, as it accumulated mass by 'capturing' space debris, through gravitational binding forces. Since then additional heat has continued to be generated by the radioactive decay of elements such as uranium, thorium, and potassium. Temperature within the Earth increases with increasing depth [3,4]. Highly viscous or partially molten rock at temperatures between 650°C and 1,200°C is postulated to exist everywhere beneath the Earth's surface at depths of 50 to 60 miles. The temperature at the Earth's centre, nearly 4,000 miles (6,400 km) deep, is estimated to be 5650 ± 600 K. The heat flow from the interior to the surface, which is the primary source of geothermal power, is only 1/20,000 as great as the energy received from the sun, but is still enough to provide a serious contribution to renewable resources, if effective extraction on a massive scale can be realised.

In assessing the role of renewables in relation to the currently popular techno-fix vision of a sustainable future it is assumed that citizens and 'electorates' in old and new industrialised nations will not choose to give up their energy profligate lifestyles by 2030 (the 'business-as-usual' (BAU) scenario). The presumption is, therefore, that the electrical supply industry will be required to match current levels of power consumption for all human activities, entirely from renewable sources. Current indicators also suggest that bio-fuels, which are already controversial, will make a very small contribution to the planets energy needs, particularly when the population of the Earth will be larger than it is now and 'land for food' will be a major issue. The crude figures for what will have to be provided are detailed below.

In 2000, the average total worldwide power consumption by the human race (population ~6.3 billion) was 13 TW (= 1.3×10^{13} W) with 86.5% from burning fossil fuels [5, 6]. This is equivalent to 3.9×10^{20} J per year, although there is at least 10% uncertainty in the world's energy consumption. Not all of the world's economies track their energy consumption assiduously. Also, the exact energy content of a barrel of oil or a ton of coal will obviously vary with quality. In 2007, given the rapid industrialisation of China and India 'on the back of abundant coal', and increasing population (75 million/year), global consumption is probably nearer 15 TW, and rising. By 2030 the population will have risen to 7.9 billion. A 1.7–1.9% increase in energy consumption [7] takes us to >20 TW in 2030. In the long term, population is predicted to level off at 10.5 billion. If we assume that energy usage per person does not fall sharply in the interim through the widespread adoption of energy conservation or through changes in lifestyles (much less mobile populations) then demand for electrical power will grow to at least 25 TW.

The 'ball-park' estimates of global energy potentially available from renewable resources when compared with the potential demands of a future energy hungry global economy, clearly seems to suggest that more than enough renewable energy exists in earthly phenomena to meet mankind's needs. The big problem is: given the short timescale of about 20 years to stop burning fossil fuels in order to bring greenhouse gas emissions down to 90–95% of 2005 levels, how much of this resource can realistically be exploited using currently available technology and how close can we get to satisfying future demand from renewables? This issue will be addressed in some detail in the ensuing sections.

3.2 Hydro-power

Where the Power Comes From

From a purely electrical engineering perspective a hydro-electric power generation plant is actually not too different from the fossil fuel power station described in Chap. 2, once we replace steam with flowing water interacting with the turbine blades. Of course, structurally and visually they are very different. As we have already demonstrated, the power in the water is furnished by gravity, provided it has been accumulated in a large deep lake or reservoir, and preferably in one which is well above sea level. The energy that can be extracted from a natural, or artificial, reservoir depends on the volume of water it contains and on the difference in height between it and the water's outflow – usually near its base for very large dams such as Aswan High dam. In mountainous schemes with many elevated natural lakes, such as the multiple loch (more poetic Scottish word for lake) schemes in Scotland, power station turbines may be a long way below the reservoirs thus enhancing the height difference. The water is usually made to fall through one or more large pipes or through a tunnel, termed a penstock, before striking the turbine blades at high velocity. This height through which the water drops is called the head.

How the Power Is Extracted

The head is critical to the design of the turbine/generator combination in any hydro-power station. The technology is very mature and design choices for getting the right turbine/generator combination to suit the conditions at a particular reservoir are well established [8]. Rotational speeds in the range 100 rpm to 600 rpm are typical of water driven prime movers, and the choice of turbine speed is governed by the priority placed on turbulence-free interaction between the flowing water and the moving turbine blades. Speed synchronisation of the streaming water and the blades is essential for high efficiency. Turbine speed is held constant by incorporating a large heavy flywheel onto the drive shaft, together with some form of controllable water flow deflectors. At limited reservoir heads in the range 10–100 ft, when water pressure is low, the turbine tends to be of the propeller type (Kaplan design – Fig. 3.1). The turbine has a compact diameter (1–5 m) for efficient conversion of relatively slow axial water flow into high rotational speed and optimised torque at the output shaft, which is connected to the generator. When there is plenty of head (more than 100 ft) the turbine is more likely to be of the water-wheel genre (water scoops on the end of radial support arms – Pelton design as shown in Fig. 3.2). While the rotating Pelton wheel itself is not greatly different in size from other types (typically 3–4 m in diameter) the water feed arrangement

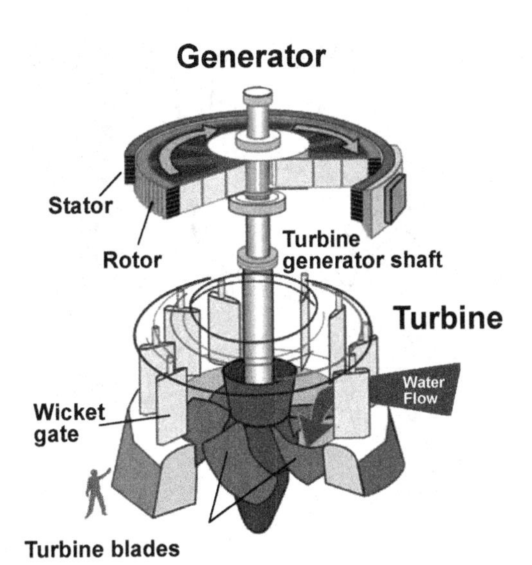

Fig. 3.1 Hydro-electric turbine exhibiting the Kaplan blade construction

to achieve efficient transference of power from the fast flowing water is complex, so that the overall diameter of the turbine can be as much as three times the diameter of the rotator. Modern water turbines are actually quite efficient in converting water power to shaft power. The value generally varies between 70 and 90% depending on precise operating conditions.

The generally slower rotational speeds available from water turbines, when compared with steam turbines, dictates that synchronous generators in hydro-electric stations are much larger than those encountered in fossil fuel power stations. In saying this it is assumed that generator outputs in the range 40 MW to 400 MW at a frequency of 50 Hz (or 60 Hz in the USA) are desired. In power stations associated with very large dams, providing potentially vast quantities of water but with a moderate head, the size of the generator means that it must be installed with the armature rotating around a vertical axis to ease bearing problems. The four generators at the Cruachan power station in Scotland, for example, which is by no means big by hydro-electric standards, are tightly housed in a cavern, which is 50 m long and about 60 m in diameter, located within Ben Cruachan. The excavation of this cavern required the removal of 220,002 m³ of rock and soil.

Although much larger than the generators typical of fossil fuel power stations, the electrical and mechanical loss mechanisms inherent in hydro-electric generators are of a similar nature to their fossil fuel counterparts and lead to similar efficiency levels of the order of 90%. Transmission losses on the grid are likely to be relatively high for hydro-electric power owing to the longer than average distances to the users from remote stations. The generally quoted figure for grid loss, as

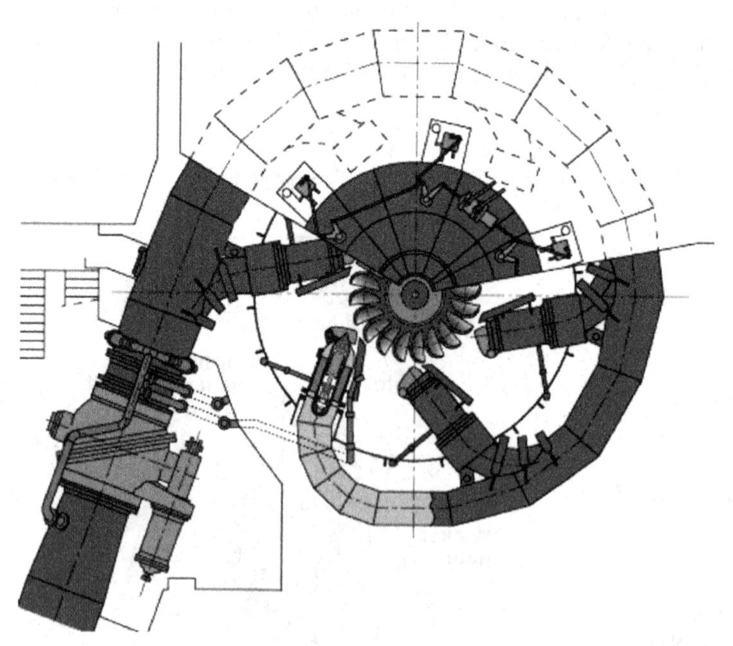

Fig. 3.2 Hydro-electric turbine employing a Pelton wheel drive mechanism

noted in Chap. 2, for all types of power station in the USA and Europe is 7%. More remote hydro-power stations will obviously incur slightly higher losses in transmitting power through the grid resulting in a loss figure of nearer 8%.

The Aswan High Dam (Fig. 3.3), which holds back the waters of Lake Nasser on the Nile, is 550 km in length, has a surface area of 5250 km^2, and contains approximately 111 km^3 of water [9]. This volume of fresh water (density = 1000 kg/m^3) has a mass of $111 \times 10^9 \times 1000 = 111 \times 10^{12}$ kg. Consequently, assuming that the average height of the water in Lake Nasser above the dam outflow is 55 m, the energy stored in the dam is $111 \times 10^{12} \times 55 \times 9.81 = 60 \times 10^{15}$ J = 60,000 TJ. However, like the pendulum discussed in Chap. 2, this potential energy yields power only when it is converted to kinetic energy. Water can be discharged at a rate of 11,000 m^3/s through the base of the Aswan dam. By performing a calculation of the kinetic energy per second (power) associated with a moving water column, a simple formula for estimating the power represented by the flow of water emerging from the dam can be constructed. The power is equal to the flow rate multiplied by the head multiplied by a conversion factor (9810 J/m^4). For a dam of this type, assuming that the in-flow is much smaller than the out-flow, the head will decrease linearly when maximum power is being extracted. The average head will be about half the maximum head. Consequently, for Aswan this gives a potential power at the maximum flow rate of 2.85 GW. Water turbine efficiencies are typically of the order of 85% while the generators are unlikely to be much better than 90% efficient, therefore of the 2.85 GW contained in the rushing water, only about 2.2 GW is available to the grid. This is very close to the claimed capability of the twelve generator sets incorporated into the Aswan High dam complex [9]. In hydro-generation systems located in mountainous terrain the head, as suggested above,

Fig. 3.3 Satellite image of the Aswan High Dam

can be greatly enhanced by arranging for the turbines to be far below the base of the dam. Gravity means that high flow rates occur at the turbine. On the other hand high confined mountain valleys are unlikely to provide very large volumes of water. In some mountainous hydro-generation schemes several reservoirs are used in cascade to raise the potential energy.

Potential as a Source of 'Green' Energy

The internet, not unsurprisingly, is a Pandora's box of much interesting information on almost any subject one can think of – not all of it reliable. Googler beware! Almost every hydro-electric power station on the surface of the globe seems to have a web site. With painstaking data tabulation from a selection of these sites it has been possible to observe that from initial planning to eventual commissioning almost all hydro-electric power stations, no matter where located, or how large or small, conform to an average gestation time of about 10–15 years. This means that with an approximately 20 year window until 2030 any large new hydro-electric power station in excess of 1 GW that has any likelihood of coming on-stream and thereby helping to replace fossil fuel usage, will have to be already substantially into the planning and approval stage of development at this point in time (2008). The World Energy Report [10] suggests that worldwide there are 77 large hydro-electric schemes (> 1 GW) at the approved or building stage with the potential to bring new renewable power into service by 2030. There are many much smaller schemes but their aggregated power is relatively insignificant in global terms. We can therefore conclude that the additional capacity which the new hydro-electric stations will bring to the generation mix by 2030 could amount to 124 GW. This is a 15% increase on current capacity. In 20 years therefore a potential total power available to the consumer from hydro-electric generation, allowing for grid and transformer losses is likely to be of the order of 840 GW, a small but significant proportion (4%) of the required ~ 20 TW.

When turbine, generator, transformer, and transmission losses for the hydro-electric system are aggregated, it is salutary to observe that in 2030, of the ~ 1.2 TW of power locked up in the streaming waters of the hydro-electric dams of the world only 0.84 TW reaches the users. A massive 360 GW disappears in heating the electrical power industry's real estate. While it is not possible to make electrical systems 100% efficient an improvement on current standards would certainly not be too difficult. In the past efficiency has never really been a pressing issue with engineers because fossil fuels were considered to be plentiful and cheap, and now renewables are often mistakenly considered to be 'free'. Each new hydro-station, although carbon clean once built, has its environmental costs. They are anything but environmentally friendly at the construction stage, if fossil fuel powered machinery is employed, while large schemes destroy farm land and disrupt the local ecology. Dams in tropical and sub-tropical regions of the world are claimed to release large volumes of methane created by decaying vegetation drowned when the reservoir was formed. So the ecological impact of medium and large dams is not insignificant. For example, stagnant water is retained in the artificial lake behind the dam and has the tendency to be under-oxygenated. The fish that live in the impoverished water that comes out of the turbines are not impressed. On the other hand, when the water from the top of the dam is suddenly released it is heavily enriched with oxygen and contains tiny air bubbles. The fish don't appreciate this either. It is not easy to keep the little blighters happy!

Improvements in efficiency could mean fewer power stations and less environmental damage. A very large proportion of the hydro-equipment in operation today will need to be modernised by 2030. This modernisation should be driven by the need to achieve efficiency improvements. Just a 1% increase in the efficiency of hydro-power stations world wide would yield a $0.01 \times 1200\,\text{GW} = 12\,\text{GW}$ reduction in electricity wastage. This is equivalent to 6–10 major new hydro-schemes, for 'free'!

3.3 Wind Power

Where the Power Comes From

A modern wind generator converts air movement into electricity. It employs a turbine linked to a generator and is, in principle, not too dissimilar to the hydro-electric arrangement described in the previous section. The turbine usually takes the form of a three bladed propeller on large wind machines in which the turbine and generator are mounted on a common horizontal axis. Three blades provide optimum stability with the fewest number of elements. However, in small wind turbines, multi-element propellers with more than three blades are not uncommon. Turbines with blades, which in shape are rather reminiscent of curved sails rotating around a vertical axis, also exist, but their wind power to mechanical power conversion efficiency is as much as 50% lower than that of the equivalent horizontal axis machine. Consequently, they have a low likelihood of being adopted by the electric power industry unless there are good non-engineering reasons for this type of installation, which over-ride efficiency considerations.

The main determinant of the power capacity of a horizontal axis wind turbine is the diameter of the blades, although their cross-sectional shape is also important [11]. The larger the diameter of the propeller the larger is the swept area through which the moving air passes. The theoretical power contained in the air stream is the kinetic energy per second passing through the swept area of the propeller. From the definition of kinetic energy given in Chap. 2, the theoretical power is equal in magnitude to the product of the air density, the swept area of the turbine blades, the air velocity cubed, all divided by two [12, 13]. However the German scientist Albert Betz (*ca* 1927) has shown that the maximum power that can be extracted from a laminar stream of air (i.e., the maximum conversion efficiency) is 16/27ths or 59% of the theoretical value. Modern aerodynamic wind turbine propellers operate with a conversion efficiency of nearer to 40%, with blade drag and air turbulence representing the main sources of this efficiency reduction. This seems low, but it is not too different from the conversion efficiency of steam turbines.

How the Power Is Extracted

The primary rotational force on the blades of a wind turbine, which aerodynamically have much in common with the wings of an aircraft, is due to 'lift' and the lift force increases with blade speed. Once the blades are rotating, velocities are high in large machines, particularly near the tips, even for moderate rotational speeds. While this is advantageous for the desired lift force, the blades become increasingly subject to 'drag'. In much the same way as a jet aircraft has thin small wings to limit 'drag' at high speed, nevertheless the wing must have sufficient aerodynamic profile and enough area to provide adequate 'lift' force at lower take-off and landing speeds. A compromise between stream-lining and satisfactory lift characteristics must be found. The same is true of wind turbine blades. In order to maximise conversion efficiency the aerodynamic lift/drag ratio must be high, and this dictates that the turbine construction tends to be more like an aircraft propeller than windmill sails which operate on the basis of drag (reaction lift) alone. The angle of attack, or 'pitch' of the blade is adjustable on medium to large size turbines to optimise the lift/drag ratio as wind speed changes. Fixed rotational rate is secured by controlling the pitch and is typically in the range 20–30 rpm for winds in the speed range 5 m/s (11.2 mph) to 25 m/s (56 mph). At 25 m/s the propeller blades are 'feathered' and the turbine is immobilised, as a protective measure. The blades are generally made from fibre-glass reinforced polyester, carbon fibre or wood epoxy. Large modern aerodynamic wind turbines capable of delivering 3 MW – not unusual for machines on North American or European wind farms – employ propeller blades that sweep an area of up to 90 m in diameter.

A turbine rotational speed of the order of 20–30 rpm is much too low for effective generator performance. In Sect. 2.5 we discovered that the AC frequency of a synchronous generator, which has to be 50 Hz or 60 Hz, is given quite simply by the number of poles times rotational speed divided by 120. At a rotational speed of ~25 rpm it is not possible to design a generator that will produce a 50/60 Hz output voltage. Consequently, a gear train has to be introduced between the turbine and the generator to raise the rotational speed at the generator drive shaft to ~1500 rpm (see Fig. 3.4). Gear trains represent a distinct disadvantage of the wind generation of electricity. They are noisy, heavy, costly, prone to wear, require regular servicing, and are a source power loss.

With an input shaft speed of the order of 1500 rpm the wind machine designer has the choice of using a synchronous generator or an induction generator. The induction generator differs from the synchronous generator described in Chap. 2, in that the armature magnetic field is set up using windings rather than permanent magnets. This makes it more tolerant to turbine speed variations – a useful feature with wind machines. At output power levels of no more than 3 MW a synchronous generator installed in a wind turbine nacelle can obviously be considerably smaller than those used in hydro and fossil fuel powered generating stations. However, turbine speed control presents a particular problem for wind turbines. Ideally the shaft speed should vary by no more than 3–4%, but this is easily exceeded even with propeller pitch control schemes. A solution that is possible for <3 MW machines involves the application of solid state electronics to decouple the electrical frequency from the rotational speed of the prime mover. It has the disadvantage of reducing generator efficiency.

Generators employed in wind systems exhibit the same type of losses as those in other types of power station, and they can be assumed to display efficiency levels that are not too different from the 90% quoted in Chap. 2. With added power electronics for frequency control the overall figure will drop to about 85%. When conversion efficiency, gear box efficiency, and generator efficiency are aggregated for large turbines, as used on wind farms, it is possible to estimate that a tenth of a square kilometre (0.04 square miles) of land (or shore) is needed by a modern wind farm to generate 1 MW of electric power to the grid. The area figure is based on turbines with ~ 100 m diameter swept areas, and the requirement for a three times diameter spacing of generating units to minimise local air turbulence. Wind farms are generally quite remote from centres of population, which means that like hydro-power we have to assume that grid losses in transmission are likely to be nearer 8% rather than the 7% normally quoted. Furthermore it is generally accepted that wind generators only deliver about 33% of capacity because the wind is intermittent. Taking these two figures, together with the land area estimate, we get the result that to deliver 1 MW to the consumer we will need of the order of one-third of a square kilometre (0.33 km^2) of global surface area. This can be converted to the rather convenient statistic of 3 W/m^2.

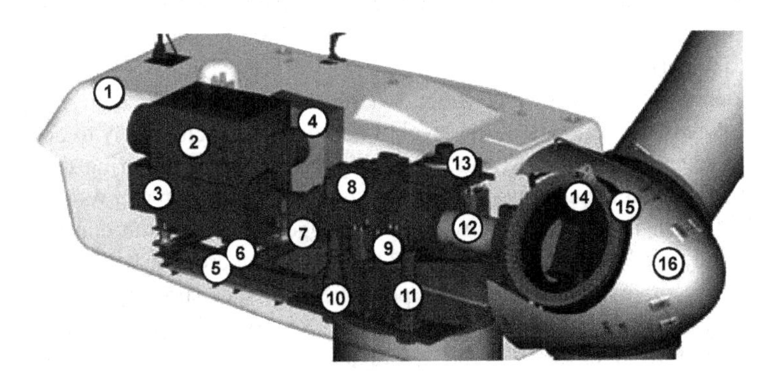

Fig. 3.4 Wind turbine schematic showing: (*1*) nacelle (*2*) heat exchanger (*3*) generator (*4*) control panel (*5*) main frame (*6*) imact noise insulation (*7*) hydraulic parking brake (*8*) gearbox (*9*) impact noise insulation (*10*) yaw drive (*11*) yaw drive (*12*) rotor shaft (*13*) oil cooler (*14*) pitch drive (*15*) rotor hub (*16*) nose cone

Potential as a Source of 'Green' Energy

So what is the maximum possible electric power that can be extracted from the wind? The web gives the land area of the Earth as 148,300,000 km^2. If we add in suitable coastal areas where turbines could be installed using current technology we get a round figure of 150,000,000 km^2. If we rule out agricultural land, and populated land, for wind farm development we have to remove 75,000,000 km^2 from the calculation. Of course Nimbyism is a strong emotion in some parts of the world but eventually, perhaps survival will be the more powerful instinct. Inaccessible mountain areas, say above 3000 ft, can also be presumed to be unsuitable as are icily cold northern and southern regions of the globe. Allowing also for ecological and environmental concerns, my guess is that this will reduce the area of the globe suitable for wind farms to an idealistic 10% of 75,000,000 km^2, which gives us 7,500,000 km^2. Finally we sensibly have to limit installations to those areas where prevailing winds prevail! Wind maps suggest that this is likely to be no more than 33% of the 7,500,000 km^2, bringing us down to 2,500,000 km^2. We are looking here at a wind farm expanse, if aggregated onto a single site, which is rather more than the area of Mexico! Finally, if the wind across 'Mexico' blows reliably, we end up with the 'broad-brush' estimate that mankind could potentially extract 7.5 TW from global winds. Seemingly therefore, a considerable proportion of mankind's energy needs can be supplied by the wind, but extracting anything like this amount by 2030 is, of course, quite another matter. It should be noted that placing wind farms on the world's continental shelves has been mooted [14]. This would raise my 7.5 TW figure to nearer 62 TW! Unfortunately to do this would require the intensive development of new deep-sea wind technology, which is far in advance of anything contained in off-shore systems that are currently deployed. The exploitation of ocean wind is not even being seriously contemplated at present, and therefore power generation from this source is certainly unlikely to happen in the next 25–30 years. Furthermore, professional engineers of the calibre and training required to 'man' such a vast and challenging project are just not being educated in sufficient numbers, even if the will to embark on such a massive enterprise were to materialise.

Having considered what could be possible, where we are actually heading is not encouraging. A realistic estimate of the level of additional capacity likely to be provided by wind generators by 2030 can reasonably be formed by filtering the copious data buried in published reports, such as the 2007 report of the World Energy Council [12]. In the section on 'wind' it is suggested that at the end of 2006 the total world wind capacity was about 72,000 MW, whereas it had been only 5,000 MW, 11 years earlier in 1995. The published statistics support the assumption that wind capacity is growing threefold every 5 years, and so we can predict, that by 2030 wind capacity could reasonably be expected to grow to 1.5 TW. It is presumed that there will be no significant worldwide government intervention, to expand qualified engineering man/power, either by intensive education programmes or by massive diversion from other activities. One or other of these options is a necessary pre-requisite for a dramatic increase in the rate of expansion of wind power production. As we have already seen power station capability has to be reduced by a third because of wind intermittency. Consequently by 2030, potentially 500 GW of wind generated electricity will be available to the grid worldwide. A 150,000 km^2 area of the planet will be required to deliver this, which would be like creating a forest of wind machines the size of the state of Illinois in the USA. While the citizens of North America might tolerate this, since wind farms are highly profitable for land owners, the general reaction elsewhere to vast expanses of turbine towers is more likely to be dismay [12]. Grid transmission losses and distribution losses will drop the 500 GW to about 430 GW available to the consumer. By 2030 environmental and ecological resistance to visual degradation on this scale may slow down development and 430 GW of useable wind capability may turn out be an over optimistic estimate. In any case assuming this figure could be reached, it represents only 0.43/20 = 2.2% of projected demand by 2030.

We will need to search elsewhere for a non-polluting source of our energy requirements. In addition, since total dependence on wind power is not possible because of its intermittency this means that an alternative reliable source of power, to 'even out' the peaks and troughs of wind, is mandatory. This could be done by ensuring that power is available from more reliable non-fossil-fuel sources, backed up by the development of more effective storage schemes than are currently available. Energy storage will be examined in Chap. 4.

3.4 Wave Power

Where the Power Comes From

In common with wind, wave power is difficult to exploit because of its diffuseness and its intermittency. But in addition, to extract energy from the waves it is necessary to go where the waves are powerful, which is also where the seas or oceans are deep and very inhospitable. Wave power can be thought of as concentrated solar power, formed when winds generated by differential heating of the atmosphere sweep over open expanses of sea or ocean transferring some of their energy into water waves. The amount of energy transferred and hence the magnitude of the resulting waves depends on the wind speed, the length of time the wind blows and the expanse of water surface over which it blows (termed the 'fetch'). In this way the original solar power levels ~ 1000 W/m^2 can be translated into ocean waves exhibiting power levels of the order of 100 MW/km of wave front as we shall see.

The nature of ocean waves is becoming increasingly well understood from studying water movement in special tanks and with the aid of sophisticated computer modelling [15, 16]. The complex motions of the water occur not just in the visible surface waves, but also well below the surface. In fact, the presence of the wave at the surface is reflected in water movements down to a depth which is of the order of about a half wavelength of the surface wave. In the deep ocean, the wavelength of the surface waves, that is the crest-to-crest distance, is approximately equal to gravitational acceleration divided by the product of twice π (3.412) and frequency squared [17]. An ocean swell exhibits a typical frequency of 0.1 Hz which gives a wavelength of 156 m. Hence the depth below which wave action is not discernable in ocean waters is of the order of 80 m. This knowledge is important to the design of effective wave energy collectors. The velocity at which the wave travels (v in m/s) is given approximately by $v = 1.25\sqrt{\lambda}$, where wavelength λ is measured in metres. Typically the crest velocity of a deep ocean wave is 16 m/s. However, the velocity expression also tells us that waves of different wavelengths travel at different speeds. The fastest waves in a storm are the ones with the longest wavelength. Observant sea watchers may have noted that when waves arrive on the coast after a storm far out to sea, the first ones to arrive are the long wavelength swells – not a bit like a high class social event!

When several wave trains are present, as is always the case in the ocean, the waves form groups which appear as higher than average ridges or pulses of water. In deep water the groups travel at a velocity (termed the group velocity) that is half of the phase speed [18]. Group velocity is associated with the energy transmission and is important in determining the power of the waves. In a wave tank it is feasible to follow a single wave in a pulse. When one does, it is possible to see the wave appearing at the back of the group, growing and finally disappearing at the front of the group. As the water depth decreases towards the coast, this will have an effect on the speed of the crest and the trough of the wave; the crest begins to move faster than the trough. This causes a phenomenon with which everyone is familiar, namely surf and breaking waves.

The power of ocean waves can be captured by devices that oscillate in response to the wave motion. The available power per unit width of regular sinusoidal waves depends on the water density ρ, gravitational acceleration g, the mean wave height H and the wavelength. It is given by a simple formula derived from energy considerations, which can be found in most textbooks dealing with water waves [16]. For strong 10 ft peak to trough, ocean swells oscillating with a frequency of typically 0.1 Hz the crude formula suggests available powers in the range of 100 kW/m (100 MW/km). Measurements from a mid-Atlantic weather station indicate that average wave power levels of 80 MW/km of wave front are potentially available there, if wave machines could be located safely and reliably, in hostile deep ocean environments. A very 'tall order' as we shall see. Nearer shore, but still in deep water, such as to the west of the Outer Hebrides of Scotland, the wave power is somewhat lower at 50 MW/km. However, it is certainly more than enough to merit examination as an exploitable resource.

How the Power Is Extracted

At the heart of any sea or ocean wave system for generating power, there is a combination of machines, whose basic function is to convert wave energy to electricity. There are essentially two system types: those that employ hydro-electric techniques and those that rely on motion sensing of a floating unit. A good example of a hydro-electric scheme is 'Wave Dragon', a system developed in Denmark [18]. An experimental 20 kW version is deployed in a North Sea fjord at Nissum Bredning. The physics underpinning it is very well known to sound engineers. The mechanism is not unlike that of an ear trumpet, which a hundred years ago was a not uncommon way of enhancing hearing. These days they more often appear in comedy sketches poking fun at senility! The horn of the trumpet guides the sound waves, from the large open end of what is essentially a metal cone, towards a narrow 'throat'. It is designed to concentrate these waves with sufficient power at the 'throat' to form a 'focused' sound wave that is strong enough to overcome the hearing loss suffered by the user. For efficient focusing, it is critical that the waves striking the horn aperture should be essentially unidirectional and that they should arrive at the device with a uniform phase front. This means that the horn cannot be too close to the sound source.

In the Danish wave collection system, which is a floating structure, reflecting 'booms' form a two dimensional horn, at the 'throat' of which is a ramp. This ramp feeds the enhanced waves towards an artificial lagoon well above sea level. The major problem with sea and ocean waves is consistent wave direction and 'good' phase fronts, and this is not helped by operation in a sea inlet or near the shore. But if a reasonable level of focusing is achieved the significantly raised wave magnitude at the ramp, will allow water to collect to a useful height in the huge floating pond. Like a hydro-electric reservoir, the artificial lagoon can potentially store large amounts of energy. Measurements [19, 20] on the Nissum Bredning prototype indicate that the efficiency in converting wave power to potential energy in the lagoon/reservoir is no more than moderate, at 30%. A low head turbine of the Kaplan type and a conventional hydro-electric generator convert the lagoon energy into electricity. Conversion efficiencies from potential energy to electrical power are no different to those of an equivalent hydro-electric system of comparable power: typically 75%. Consequently, ocean wave power stations of this type are unlikely to extract much more than 20% of the energy in the waves.

Motion type systems for extracting wave power are designed with two basic requirements in mind. First, the energy gathering mechanism, associated with the motion of the sea, must be isolated, as far as possible, from the mechanism for converting the extracted mechanical power to electrical power. This maximises system reliability in the harsh marine environment. A second requirement is that the principle of operation must be very robust and the machines designed to implement it must be capable of withstanding the most severe battering from ocean storms. Again there are two types: wide structures aligned at right angles to the incident wave direction (terminators), and long thin usually floating structures aligned in the direction of travel of the waves (attenuators). Examples of terminators that have been proposed in recent years, are the Salter duck, the Cockerel raft, the clam, and the oscillating water column (OWC). Of these, only systems based on the oscillating water column concept have come close to delivering commercial levels of power output. Machines that are not at an advanced prototype testing stage by 2008 are unlikely to be delivering real power in significant quantities by 2030. So I shall concentrate on examining only OWC operation and efficiency.

The OWC is based on the concept that an entrained column of water (Fig. 3.5), within a fixed and partially submerged open-ended concrete chamber, will be excited into motion by wave action, and will as a result act as a massive piston. As the trapped water column surface moves up and down it will pump large volumes of air in the chamber above it. For example, a proposed ocean wave farm at Mutriku in Northern Spain will comprise eight concrete water column chambers each 5.5 m in diameter and 3.1 m deep. The anticipated capacity of this system is 0.3 GW. This is small by hydro-electric and wind power standards but is significant for the, as yet, immature wave generation industry. The compressed air above the water column is allowed to escape at high velocity through an aperture at the top of the chamber towards an air turbine and generator. Air is also drawn through the turbine as the water column falls and 'rectifying' turbines have been designed to take advantage of this. The turbines proposed for the Mutriku farm are designed [21, 22] to realise a capture factor of 60%. That is, 60% of the power in the waves will be converted to air pressure in the turbines. Air turbines suitable for the hostile marine environment [22] are typically 60% efficient in converting the power in the moving air into shaft power at the generator. Induction generators are the best choice for this application and, as we have seen, these machines are capable of 90–95% efficiency. Inverters for the rectifying turbines add 4% to the system losses. Consequently, we are looking at power plants of the OWC type being capable of converting about 33% of the power in the waves into electrical power to the grid.

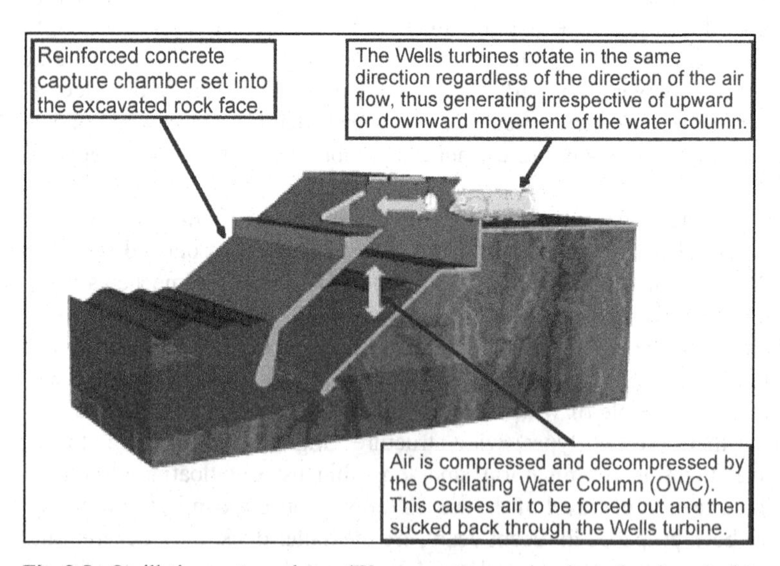

Fig. 3.5 Oscillating water column (Wavegen concept drawing) showing air driven turbine at top right

Wave power systems of the attenuator genre include tethered air bag designs in which wave action results in air pressure changes in flexible floats, or rigid floats joined together to form a long articulated structure, which flexes in sympathy with the waves. Of these, only systems of the latter category have advanced to a prototype stage. The best known system is termed Pelamis, which is composed of a series of cylindrical hollow steel segments that are connected to each other by hinged joints [23]. The device is approximately 120 m long and the cylinders are 3.5 m in diameter. It is tethered to the sea floor in approximately 50 m deep water so that the structure remains aligned at right angles to the wave fronts. As the waves progress down the length of the structure it 'ripples' in a snake like fashion and the relative movements are picked up by hydraulic pistons at the hinged joints. The hydraulic actuators pump oil to a hydraulic motor/generator set via an energy smoothing system. The capture efficiency in converting wave power to hydraulic power is comparable with the OWC at about 60%, while the hydraulic motors typically exhibit efficiencies of the order of 85%. With a 90% efficiency for the electrics – asynchronous generator, plus 11 kV to 33 kV transformer – the efficiency in converting wave power to grid power for a Pelamis system is estimated to be about 45%. Nevertheless, it is enough to have encouraged Portugal into developing the world's first commercial wave farm, at the Aguçadora Wave Park near Póvoa de Varzim. The farm will initially use three Pelamis P-750 machines generating 2.25 MW. Subject to successful operation, a major investment is planned by 2009 for a further 28 machines building to a capacity of 525 MW.

Potential as a Source of 'Green' Energy

Good wave power locations ideally exhibit a flux of about 50 kW/m of shoreline. Variable sea conditions suggest that capturing 20% of this (or 10 kW/m) in sites that are not unfeasibly hostile is within the realms of possibility. Assuming very large scale deployment of (and investment in) wave power technology, coverage of 5000 km of shoreline (worldwide) is plausible [22], although not by 2030. Therefore, the potential for shoreline-based wave power is about 50 GW. Given that only half of this power can be converted to electrical power, because of wave system inefficiencies, we can assume that global wave power can, at best, provide 25 GW to the grid and 22 GW to the consumer. This is a tiny fraction (0.11%) of the 20 TW likely to be required by mankind by 2030. Deep water wave power resources are truly enormous, but can be completely ruled out as far as major exploitation in a 20–25 year time frame is concerned.

The difficulties of exploitation in deep sea environments are of such a severity that it has frightened off investment in marine solutions and this has resulted in wave power being the laggard of the renewables industry. Even with a massive turn around in investment today (i.e., in 2008), the wave power contribution to mankind's needs by 2030 will remain relatively insignificant.

3.5 Tidal Power

Where the Power Comes From

Tidal flows, as we have seen earlier, result from the gravitational influences of the sun and moon acting in concert. The rise and fall of the sea or ocean level varies approximately sinusoidally with a period of 12.4 hours. This is termed the diurnal cycle. It is superimposed on a longer cycle – the spring/neap cycle – of 14.7 days or 353 hours. Spring tides are those tides that have maximum amplitude (above average sea level), and they occur when the moon and the sun are aligned and their gravitation pulls are additive. At neap tides, minimum tidal amplitude occurs, when direction lines from the sun to the Earth, and from the Earth to the moon, form a right angle. The ratio between the amplitudes of the maximum spring tide and the minimum neap tide can be as much as a factor of three. Smaller seasonal variations also occur. The tidal variation from high tide to low tide is referred to as the tidal range. In a fictitious Earth covered in ocean of constant depth the range would be about 1 m, but in practice it is amplified in coastal areas by complex interaction with coastal terrain and other effects of the rotating globe. The greatest tidal magnitude occurs in estuaries where the advancing tidal wave – the flood tide – is reinforced by secondary waves reflecting off the walls of the estuary forming an amplified stream of water. By using the potential energy formula given in Sect. 3.1 applied to the fictitious Earth's ocean with a 1 m tidal range, we can estimate that tidal energy stored in the seas and oceans is of the order of 7,000,000 TJ. This is equivalent to 130 Aswan dams. Not a lot in global terms, but not insignificant. The problem is that the tidal energy trapped in the oceans and seas is difficult to access.

How the Power Is Extracted

Power can be extracted from the tide by two means [24,25]. First, by building what civil engineers would term impoundment ponds, or basins, so that as the tide rises water is channelled into the artificial reservoir through a turbine and released as the tide falls through the same turbine; or second, by inserting a water turbine into a tidal stream wherever such natural phenomena exist. Impoundment dams across an estuary or firth are expensive to construct, the natural water cycles are completely disrupted, as is navigation in the estuary. However, with multiple impoundment ponds power can be generated to match consumer demand. On the other hand, with tidal stream systems, the generation is entirely determined by the timing and magnitude of the currents. Unfortunately, the best currents may be unavailable for power generation, because the turbines and generator housing structure would obstruct essential navigation through the strait or channel giving rise to the tidal stream.

Extracting energy from the tide by building a dam (barrage) across an estuary or coastal inlet is in principle the simplest solution. A basin or reservoir is thus formed behind the dam, which incorporates turbine/generator sets to generate electricity. These sets are not greatly different to those employed in low-head hydro-electric systems. In its most basic form the rising flood tide enters the basin through gated openings (sluices) in the dam and through the turbines usually idling in reverse. When the rising tide reaches its peak all openings are closed until the sea level has ebbed sufficiently to develop a usable head across the barrage. The turbines are then opened to allow the water collected in the reservoir to flow back into the sea. Electricity is generated for several hours until the difference in water level between the emptying basin and the next flood tide has dropped to a point where the difference between the basin level and sea level is insufficient to power the turbines. Shortly afterwards the levels will be equal, the sluices are opened and the cycle repeats. Studies have shown that this method of operation, using ebb flow only, results in the lowest unit cost of energy particularly if combined with pumping at high tide. With ebb generation from a single basin, electricity is produced for 5–6 hours during spring tides and about 3 hours during neap tides out of a tidal cycle lasting approximately 12.4 hours; thus a tidal barrage produces two blocks of energy each day, the size and timing of which follows the lunar cycle. This restriction is eased if more than one reservoir is available. As the timing of the tides and hence the generation period drifts backwards each day by about one hour, the generation (and pumping if used) need to be planned in advance to integrate with consumer demand and supply to the grid.

The average power output from a tidal barrage is approximately proportional to the square of the tidal range with the energy output approximately proportional to the area of the water trapped in the barrage. As a guide to judging the economic feasibility of power generation from a barrage the minimum mean tidal range should be at least 5 m. Assessments of technical and economic feasibility of tidal barrages are site specific. Some locations are particularly favourable for large tidal schemes because of the focusing and concentrating effect obtained by the bays or estuaries. The largest scheme currently in operation is the 40 MW barrage at St Malo in the La Rance Estuary of France, producing some 500 GWh per annum. Work on the La Rance site commenced in June 1960, with the final closure against the sea occurring in July 1963. The last of the twenty-four 10 MW turbines was commissioned in November 1967. The La Rance system consists of a dam 330 m long, forming a basin of 22 km^2 surface area, with a tidal range of 8 m.

Relatively rapid marine currents occur in constraining channels such as occur in straits between islands, shallows between open seas and around the ends of headlands. Marine currents are driven primarily by the tides, but also to a lesser extent by coriolis forces due to the Earth's rotation, salinity and temperature differences between sea areas. Typical velocities at peak spring tides are in the region of 2 to 3 m/s or more. The main requirements for their exploitation for power generation are:

- fast flowing water;
- a relatively uniform seabed to minimise turbulence;

- sufficient depth of water to allow large enough turbines to be installed;
- above conditions extending over as wide an area as possible to allow the installation of enough turbines to make the project cost effective;
- free from shipping constraints; and
- near enough to a shore-based electricity supply network capable of taking the power delivered.

The extraction of energy from marine currents by means of propeller-like turbine rotors (Fig. 3.6) is governed by the same equations as for wind turbines. Thus the power available from a stream of water through a turbine is equal to half the product of the water density times the area swept by the turbine blade times the stream velocity cubed [17]. The power per unit area delivered by flowing water is, however, much larger than for the wind. While the density for salt water is 1030 kg/m^3, the density of air at 10°C is 1.2473 kg/m^3. Consequently, for stream or wind velocities of a similar magnitude, say 3 m/s, the power density in the sea stream is almost 70 times that of the wind. Despite this apparently huge advantage, at the school physics level of calculation I hasten to add, tidal stream developments are in their infancy because exploitation is complicated by the fact that the natural stream in a geologically formed sea channel, can be completely disrupted by the extraction equipment, if the site is not carefully selected [26]. Few sites, it seems, meet the necessary conditions. Prototype tidal stream systems (Fig. 3.6) have been installed in the Strait of Messina (between Sicily and Italy), in the Bristol Channel in the south-west of England, and in the Strangford Narrows (Northern Ireland). In the USA a system is being developed for installation in New York's East River. Also, New Zealand has particularly interesting opportunities, because the tidal pattern results in a state of high water moving around the country

Fig. 3.6 Illustration of SeaGen, a tidal stream turbine generator (Courtesy of Lir Environmental Research Ltd.)

once per tidal cycle. It is postulated that with well sited tidal barrages, or tidal stream systems, electrical supply to the country could be decoupled from the natural lunar cycle. The location of power stations at Manukau harbour and Waitemata harbour, for example, which are relatively equidistant from Auckland, would lessen power supply variations to the city.

Potential as a Source of 'Green' Energy

The indications are that there is a ceiling to tidal power, and calculations suggest that it is of the order of 0.4–0.5 TW. Furthermore it seems that a large fraction of this is unexploitable. Suitable sites for barrage and tidal stream systems are scarce. In the long term future it seems just possible that 0.2 TW could be extracted from this resource. According to the World Energy Council [26] report of 2007, a rather optimistic 0.16 TW of installed capacity is being planned around the world. However, it is clear that by 2030 tidal power will not form a significant part of the renewable energy mix that will be required to counteract global warming trends.

3.6 Solar Power

Where the Power Comes From

The current estimate for the solar constant is 1367 W/m^2. At the beginning of this chapter it is noted that the solar constant is a measure of the radiant power from the sun that is intercepted by the Earth's disc. However, the surface area of the Earth is four times that of its disc [27], and if we assume that the radiant power is reduced by half both through reflection at the boundary of the Earth's atmosphere and space, and in absorption by passing through the atmosphere, we have to divide the solar constant by eight to get the figure for solar power, sometimes termed irradiance (curiously the term 'insolation' is preferred in USA scientific literature), at the Earth's surface. On average [28] it exhibits a magnitude of about 170 W/m^2. In fact in hot equatorial areas such as Arabia it can be as high as 300 W/m^2, while in cloudy Northern Europe it is nearer 100 W/m^2. The important question is: how much of this power can be a reliable source of electricity by 2030 using currently available technologies? This, of course, depends on how efficiently it can be converted to electrical power for use by consumers.

How the Power Is Extracted

Several technologies are actively being pursued with the aim of exploiting solar radiation: photovoltaic (PV) methods, solar thermal electric or concentrated solar power (CSP) techniques, passive solar design (PSD) and active solar. However, since only photovoltaic and solar thermal electric methods have the object of generating electricity, it is clear that our attention should be directed toward these technologies.

Photovoltaic technology has, rather stealthily it seems, been becoming more and more of a pervasive influence in modern consumer electronics since the 1980s. Solar cells are now common in calculators, watches, radios and toys, and are increasingly to be found powering street signs, parking meters and traffic lights. Like most technologies that sustain modern ways of living, most people accept it but do not understand it. For students of electrical science PV can actually be a difficult subject because it is usually taught from a quantum mechanical perspective. However, the PV effect can be understood well enough to make competent decisions about how it should be used by employing a crude but rather effective gravitational analogy of a PV junction, much as we have done in the previous chapter to explain other electrical phenomena. You may remember that we likened the passage of electrons along a conducting wire to balls rolling down a pin ball machine. With a little bit of imagination we can expand this analogy to illuminate PV action. It is worth noting that at lower than light frequencies the phenomenon is still present, but is interpreted as semiconductor diode detection.

Semiconductors are primarily formed from silicon. It has a crystalline structure in which the atoms bind together by a generous and powerful sharing of electrons. In its pure form it is a poor conductor at normal temperatures, since few of these electrons are 'free' as in a conductor. However, when silicon is doped with a small amount of arsenic, atoms of the dopant get bound up in the silicon crystal lattice. But arsenic has one more electron in its atom than silicon and this electron does not get used in the binding and sharing process. Consequently the impurity atoms contribute 'free' electrons which can drift through the crystal. The material, termed N-type silicon, now conducts although not as well as a metal. A similar process occurs when silicon is doped with a material like indium, which has one fewer electron in its atom. In this case at the positions within the silicon lattice where the indium has taken the place of a silicon atom a binding electron is missing, forming a 'hole' in the lattice. Any free electron entering this P-type material will be attracted into the 'holes' and conduction again results. However, the really interesting aspect of this silicon doping exercise is when N-type material is in contact with P-type material [29].

At the instant when the N-type and P-type materials are brought together, the second law of thermodynamics comes into play encouraging the 'free' electrons in the N-type material to diffuse across the junction, additionally so, because of the P-type holes that are waiting to be filled. The process of diffusion is quite common in nature; it is picturesquely present when a smoke ring spreads inexorably into the still surrounding air, dispersing in accordance with the second law. In the PN junction the process will continue until the negative charge on the P-side of the junction and the positive charge (due to electron deficiency) on the N-side result in a voltage across the junction, and hence an electric field, which prevents further charge movement. This electric field is a charge separation field, as described in

Chap. 2. The gravitational analogy would be a pin ball machine with two levels and a ramp where they join up. At the upper level (N-side) there are pins and a number of energetic balls reverberating around. On the lower (P-side) there is a similar number of pins, a few holes just large enough for balls to fall into and far fewer pin balls than on the upper level. On a flat table by dint of fortuitous collisions in the upper tier a pin ball may occasionally head towards the ramp and drop to the lower layer. Some of these will disappear into holes, but those that do not, and continue to bounce around on the lower layer will not get back to the upper layer because of the ramp. At this point, the analogy only partially describes the semi-conductor diode action, because there is nothing to stop the rattling pin balls on the upper level continuing to reach the lower level. This can be corrected by introducing a gravitational field to model the electric field in the diode. We need to imagine that balls falling into the holes trigger a mechanism that tilts the table, raising the lower end, and lowering the upper end, until the point is reached where pin balls heading for the ramp are turned back by the slope of the table, i.e., by the force of gravity. The wave detection analogy is almost complete. Consider, finally, what happens if the table is rocked very gently about this stable state. Increasing the tilt will have virtually no effect on ball movement down the ramp, with the increased gravitational force further discouraging even the most energetic balls from approaching the ramp. It is assumed that the tilt is never enough to allow balls to roll up the ramp. On the half cycle of the rocking movement when the tilt is reduced towards zero we return to a state where gravity is again insufficient to prevent some balls on the upper level finding their way over the ramp. Thus the oscillating movement results in a one way current of balls (DC when averaged) on the tilt lowering half cycles. Crudely the rocking table (AC) produces one way ball flow (DC). This is AC to DC conversion.

AC–DC conversion is also the process that occurs in a semiconductor junction immersed in electromagnetic waves. The action of the electric field of the wave on the electrons in the semiconductor junction layers is not unlike the effect of the tilting table on the pin balls. When the electric field across the junction due to the electromagnetic wave, is in the direction of reducing the charge separation field, electrons will start to find their way across the junction, and a current flows (see Fig. 3.7). On the other hand, in the half cycle of the wave when the electric field enhances the charge separation field, electrons continue to be prevented from crossing the junction. The averaged charge flow across the junction thus contributes to a DC current through the semiconductor diode resulting from its immersion in the AC electromagnetic wave. At light frequencies the process is more complicated owing to quantum effects and photon absorption, which enhances the current generation mechanism. In fact modern solar cells actually have a thin layer of intrinsic material (undoped silicon) between the P-type and N-type semiconductors (PIN diodes), which helps improve photon collection and hence efficiency. A solar cell's energy conversion efficiency is defined as the percentage of power converted from absorbed light into electrical power, when it is connected to an electrical circuit. The DC power generated by a solar cell is given approximately by the product of its efficiency, its area in square metres, and the irradiance

[30, 31]. A 0.01 m^2 cell with an optimistic efficiency of 20%, located in the Sahara desert and pointing at the sun with an irradiance of 300 W/m^2 will generate about 0.6 W. Of course the sun shines for only 50% of the day in near equatorial regions so we have to assume that we will collect 0.3 W averaged over time. Clearly we will need an awful lot of cells of this area to generate significant power.

Most solar cells in operation today are single crystal silicon cells. The silicon is purified and refined by well established techniques into a single crystal (typically 120 mm in diameter and 2 mm thick) and then micromachined to form 50 μm thick wafers [32]. (A micrometer (μm) is one thousandth of a millimetre (mm).) The technique involves taking the silicon crystal, and making a multitude of parallel transverse slices across the wafer (rather like finely slicing a round loaf of bread) creating a large number of wafers, which are then aligned edge to edge (slices of bread laid out flat to be toasted in the sun) to form a cell comprising 1000 wafers of dimensions 100 mm × 2 mm × 0.1 mm laid end-to-end on the 100 μm edges. A total exposed silicon surface area of about 2000 cm^2 per side is thus realised. As a result of this slicing, the electrical doping and contacts that were on the face of the original crystal are located on the edges of the wafer, rather than the front and rear as is the case with conventional cells. This has the interesting effect of making the cell sensitive from both the front and rear (a property known as bi-faciality) [32]. Using this technique, one silicon crystal can result in a cell capable of generating 10–12 W of electrical power in bright sunlight. In order to achieve this level of power output from un-sliced silicon crystal cells we would require about 40 crystals. The electrical contacts formed from evenly spaced metal tabs on the wafer edges are connected (an added technical complication in the fabrication of sliced silicon cells) to a larger 'bus' con-

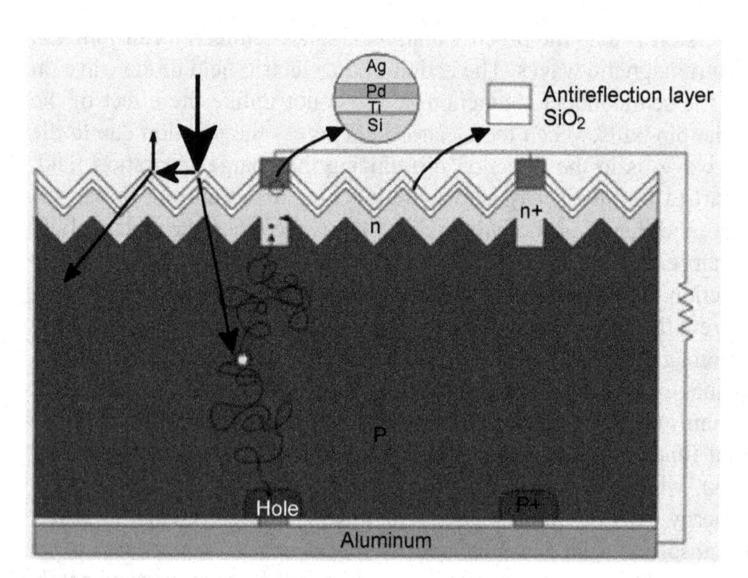

Fig. 3.7 Schematic of photovoltaic crystal showing an electron dislodged by photons striking the surface drifting though the P-layer

ductor to transmit the power. The cell is covered with a thin protective layer of dielectric with an anti-reflective coating. Cells of this construction generally represent a good compromise between cost effectiveness, reliability and efficiency. In very large solar array systems [33] cells are incorporated into panels that are approximately $3\,\text{m} \times 3\,\text{m}$ in area, each capable of producing a power of about 260 W averaged over time. Several panels (typically 100) are combined to form a module delivering something like 25 kW and will occupy a ground area of $320\,\text{m} \times 3.2\,\text{m}$ when stiffening frames, expansion gaps and windage gaps are factored into the area calculation. A 100 MW sub-array would typically be formed from 4×1000 modules and will expand the ground coverage to $1.3\,\text{km} \times 3.2\,\text{km}$ (ignoring support structure and access spacing). To get to a decent sized solar array power station we need 10 (2×5 say) of these. A ground coverage of up to $2.6\,\text{km} \times 16\,\text{km}$ results. The expanse of ground required to accommodate such an array, has to be almost four times the area of the array itself, to allow space for structural supporting frames, plinths, access roadways, and automatic cleaning systems, and to locate transmission lines, inverters and transformers – so we are looking at 160 million square metres to produce 1 GW at the array. The solar power conversion factor per unit area of global surface reduces to $6.3\,\text{W/m}^2$, from the $170\,\text{W/m}^2$ irradiance level often used rather too optimistically in documents advocating the merits of solar power.

The electrical connections between cells, between panels, between modules and between sub-arrays are designed to achieve a voltage in the range of about 6–7 kV. This is, of course, DC and it is necessary to convert this to 50/60 Hz three phase AC, at a voltage of about 500 kV for long range transmission across the conventional grid. DC/AC conversion is technically quite simple, and essentially involves switching the direction of the DC current in the primary winding of a step-up transformer. The arrangement is termed an inverter. In high power, high voltage systems, solid state thyristors, or mercury arc valves [34] can be used to perform the electrical switching. With the help of harmonic filtering, a three phase sinusoidal AC voltage, at the level of the grid is thus formed across the output terminals of the secondary of the transformer. To handle an array power level of 1 GW, the inverter system is large and is likely to require a power house occupying an area of about $300\,\text{m} \times 300\,\text{m}$, on the solar farm site. When switching losses, transformer losses, and mismatch losses within the power house equipment, and optical degradation losses in the array, are taken into account, power to the grid is diminished by a factor of about 0.78. Therefore to launch 1 GW on to the grid we require approximately 200 million square metres of desert landscape, and this equates to an area conversion factor for solar power of $5.0\,\text{W/m}^2$.

The alternative way of converting solar power to useful electricity employs much more conventional technology. The concept underpinning concentrated solar power (CSP) could be described as child's play! Many children, at some stage in their play activity, are likely to have discovered, or been shown, that a magnifying glass creates a bright hot spot on paper, which has sufficient power density to cause the paper to singe and hence to etch a hole. Every scout used to know that this was the only legitimate way to start a fire! Match-sticks were cheating. The magnifying glass if properly shaped concentrates the parallel rays of the sun by bending (remember Snell's laws [35] from school physics?) them through the lens and directing them towards a focus, where the paper should be located. In very large scale CSP systems lenses would be far too expensive and much too cumbersome and heavy to distribute over many square miles of desert, so instead ray concentration is achieved using curved moulded reflectors. These systems are really the inverse of a car headlight on a very large scale. If a car headlight reflector were used to collect the rays of the sun on a bright sunny day a hot spot of light would be formed where the bulb is normally located. In panel sizes appropriate for the forming of large solar arrays, parabolic reflectors are relatively inexpensive, they are not heavy, and, importantly, they can be manoeuvred electronically to track the sun. The requirement to focus the concentrated solar energy on collectors and to continue do this as the sun traverses the sky, means that CSP farms must be located on stable terrain. In addition they are restricted to land areas where winds are generally light to ensure minimal disturbance to the alignment of the optical reflectors.

The technology of CSP farms comprises the following six basic elements: a collector, a receiver, a fluid transporter, an energy convertor, a generator and a transformer. All of these sub-systems can be realised today using well established and available technology. Needless to say a range of system topologies are under development each of which has its advantages and disadvantages. The alternative arrangements are essentially distinguished by the way in which the solar reflectors are organised to concentrate the light onto a receiver containing a working fluid. In parabolic trough systems the reflectors (curved in one plane only – see Fig 3.8) are arranged in parallel rows (usually in north–south alignment) directing light onto long straight receiver pipes lying along the focal line of the

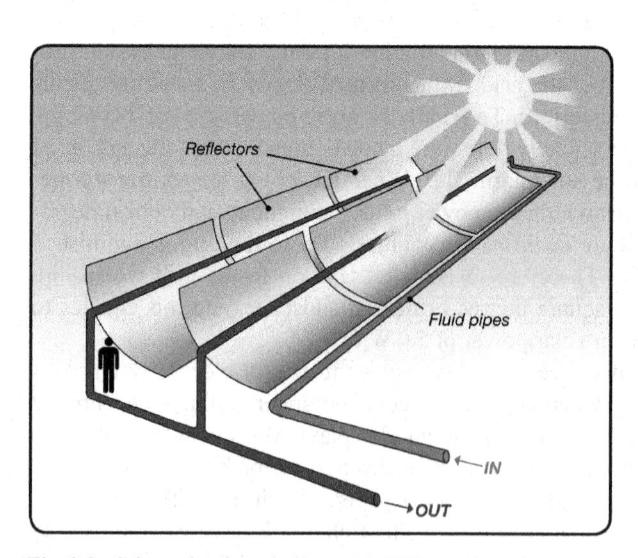

Fig. 3.8 Schematic of parabolic trough CSP system

trough. The changing height of the sun in the sky as the day progresses is accommodated by a tracking system, which very slowly rotates the mirrors about a horizontal axis. Fluid flowing, under pressure, through the receiver tube is heated to between 100 and 500°C, and then transported through a well insulated network of pipes to a boiler, to generate steam. The conversion efficiency from solar power incident on the reflectors to heat in the receiver fluid is of the order of 60%. Included in this figure is 91% reflectivity for the mirrors and 95% interception by the receiver. The superheated steam system can be expected to perform to an efficiency of about 85% [36]. Thereafter the solar plant is not greatly different from a coal-fired electricity generator, with the steam feeding a conventional turbine (efficiency = 40%), followed by a synchronous generator with an efficiency of 90% and a step-up transformer (efficiency = 95%). These figures give a ball-park estimate for conversion efficiency, from solar incident power to electrical power to the grid, of 18%. To generate 1 GW at the grid, on a hot arid desert with 300 W/m^2 of irradiance for 50% of the day, we will need 37 million square metres of reflectors. Given that real estate required to accommodate this area of reflectors is close to five times reflector expanse (estimated from studying Solar Energy Generating Systems (SEGS) in the Mojave desert [37]) then we need 185 million square metres (5.4 W/m^2). Within the error range implicit in the way the above efficiency figures are estimated, it is reasonably valid to assert that the trough CSP system and the PV system are largely comparable in performance, in relation to their overall conversion efficiency of solar power to grid electrical power. It is hardly surprising that this should be so. Otherwise competition between the two systems for major funding contracts would not be so fierce.

The other CSP formats that have been proposed envisage mirror arrangements that provide higher optical power density at the focus of the reflectors. In the so-called heliostat system the individual parabolic reflectors are arranged in rings around a central tower. It is claimed to have two basic advantages over the trough system. First, the sun can be tracked in both elevation and azimuth, and second, the fluid passing through the receiver on the central tower is raised to a much higher temperature in the range 800°C to 1000°C. This promises greater efficiency, although no full scale prototypes have been built to establish this. Consequently, it seems reasonable to conclude that this system is too early in its development to be considered to be a contender for major deployment in the deserts of the world by 2030.

A third system, which is also at the early prototype stage of development, is based on solar ray focusing by a circular (~ 40 ft diameter) dish-shaped parabolic reflector, each of which with its receiver is a stand alone electricity generator. Power station levels of electrical power are gathered from large numbers of these deployed in an extensive regular grid in a suitable desert scenario. Several prototype installations of limited size have been operating successfully over the past decade. Each dish is like a very large car headlight reflector and all are automatically controlled to accurately focus the suns rays on to the receiver. The sun is, again, tracked by tilting the dish in both elevation and azimuth. Despite the additional complexity and manufacturing cost of the large dish-shaped parabolic reflectors and their sophisticated support structures, the arrangement has two distinct advantages. First, the system is modular, in so far as every dish and receiver set is an independent solar power station (rather like a wind generator) and consequently they can be installed and efficiently operated on hilly terrain, unlike trough and heliostat systems. Second, by replacing the fluid mechanism for transporting the heat generated by the focused solar rays, with a device in the focus of each dish that converts the solar heat directly into electricity, efficiency improvements can be realised. This device comprises a Stirling engine coupled to an induction generator. The Stirling option becomes feasible with operating temperatures in the region of 700°C.

The Stirling engine is in many ways much like the petrol or diesel engine that powers your car, except for one major difference. It is an external combustion engine rather than an internal combustion engine. While the gas in the cylinder of a petrol engine (vaporised petrol) is ignited by a spark and burnt internally, and in a diesel engine the vaporised diesel is ignited internally by pressure then burnt within the cylinder, the working gas in a Stirling engine, usually hydrogen, is sealed into the cylinder and is not burnt. Piston movement is caused by thermal expansion of the gas by the external application of heat through a heat-exchanging interface material. In the solar dish type array the heat is supplied by the focused sunlight. At peak operation (irradiance greater than 250 W/m^2) the conversion efficiency from solar power collected by the parabolic dish to electrical power supplied by the generator is claimed to be 30% on the basis of prolonged testing [38]. The system efficiency is, however, susceptible to daytime irradiance dropping below the optimum level due to clouds or haze, and this means that over time, a 23% conversion efficiency for solar power to electrical power to the grid, for a large farm of this type, is more representative of its real capability. This is still better than trough and heliostat systems because of the avoidance of the losses associated with the inefficient transfer of power to the steam turbines through the agency of a hot fluid.

It must be clear to anyone who has driven a car with windscreen wipers on a very slow sweep that the presence of moisture, or rain drops, on the screen distorts and attenuates forward vision. The same is true of the optical surfaces of a PV array, or on the mirrors of solar concentrators. Optical distortion of this description can be very deleterious to solar array efficiency. Consequently, large solar power stations are planned [33] to be sited in desert areas, where solar irradiance is high and optical contamination, and therefore optical distortion, is minimised in the dry atmosphere. The most suitable arid desert locations [33] are the Sahara (8.6 million square kilometres), the Gobi (1.3 million square kilometres), the Thar (India: 0.2 million square kilometres), the Negev (Arabia: 0.001 million square kilometres), the Sonoran (Mexico: 0.31 million square kilometres), the Mojave (California: 0.7 million square metres) and the Great Sandy (Australia: 0.4 million square kilometres), giving an area of 11.5 million square kilometres, although not all of this is sufficiently flat to accommodate vast trough arrays or sufficiently wind-free to keep possible mirror misalignments to a minimum. The total desert area for the planet is closer to 17 million square kilometres, but many of the other deserts such as the Great Basin and the Chihuahuan in Mexico lack a sufficiently dry, cloud free climate, and lack suitably large expanses of level ground, to provide attractive solar 'farming'.

The deserts identified above, and solar power stations if located there, will be quite remote from the communities which they are intended to serve. For example there are quite advanced plans to supply the future electricity needs of Europe from solar farms on the Sahara desert [33] but this entails very long transmission distances – 3000 km and more, with submarine cables carrying electrical power from North Africa to Europe on the floor of the Mediterranean Sea. As we have already noted in Sect. 2.6, very long overhead power lines, and particularly long underground or undersea cables, present significant transmission difficulties for AC systems because of reactive power loss. It is necessary to introduce frequent shunt compensation along the cables to minimise loss and stability problems. These interconnections increase fault occurrence levels for the overall system. The solution, as we observed in Sect. 2.6, is DC transmission, which suffers none of these difficulties. The 'spin' attached to HVDC (high voltage direct current) is that losses are much lower and installation costs are less than for AC, but it is important that we are clear what is meant by 'losses'. While reactive power losses are no longer a problem using HVDC, ohmic or joule heating losses continue to feature. In Sect. 2.6 we noted that these losses contribute 8% per thousand kilometres for an AC line. The disappearance of skin effect for DC transmission means that this figure is reduced to about 6% per thousand kilometres. Distances from solar farms to consumers are much greater than those associated with the current grid system, so that on average transmission losses will be nearer 15% than 6%. A further 1–2% is lost in distribution, which means that when transmission and distribution losses are factored into the solar power to consumer equation, we end up with a figure of 4.5 W/m^2.

Potential as a Source of 'Green' Energy

With a conversion rate for solar power to the consumer of 4.5 W/m^2, it is theoretically possible, using an area of 11.5 million square kilometres representing the area of identified suitable deserts, to extract from solar power 52 TW; three times current global power consumption. Of course, it is pure fantasy to consider the deserts of the world being covered completely by solar farms. Even deserts support rich and diverse forms of life and have environmental and ecological importance [39, 40]. This would clearly be jeopardised if blanket coverage by solar farms were perpetrated on them, although it is not difficult to find in the energy industry literature, 'artist's illustrations' attempting to depict what tens of millions of acres of Arizona desert could look like, if covered in optical reflectors. So clearly it is not an impossible concept for some. Concern for desert peoples, such as aborigines in the Great Sandy, 2.5 million nomads in the Sahara, nomadic Mongols, Uyghurs, Kazakhs in the Gobi, and for desert ecology, mean that it will not be wise, sensible or prudent to allow solar farms to cover more than 8% of desert land area [33], simply replacing one form of pollution with another, without being fully cognizant of the possible environmental impact. 8% is right at the top end of what is considered to be within the bounds of possibility, given time and manpower. To put this in perspective – we are talking about covering an expanse of the globe with solar panels, equal in area to France plus Spain plus Portugal! Limitations will also be created by the vulnerability of vast farms to encroachment by unfriendly human beings (en masse the species is remarkably unintelligent and warlike) intent on mayhem, leading to intractable security and protection issues, to severe maintenance difficulties, and to major safety and reliability concerns. Assuming humanity is prepared to tolerate, and could protect and maintain, solar farms spread over a dispersed area of the proportions of France plus Iberia, the conclusion emerges that such farms could, in principle, generate a grand total of 4.2 TW of electrical power to the consumer, but clearly not by 2030, without an unprecedented redirection of financial resources to make it happen. Small scale solar systems for local heating and lighting could perhaps add about 0.3 TW to this giving a long term goal for solar of approximately 4.5 TW; 28% of current global needs. This solar contribution of 4.5 TW to the global demand for energy is clearly in the realms of the possible at some point in the future, but what is achievable by 2030, on the basis of currently incoherent energy policies? A range of sources of statistical data exist in which growth trends for the installed capacity of solar power stations are presented. Unfortunately the predicted rates often seem to be linked to the agenda of the sponsor of the report. Estimates for global solar capacity in 2050 can differ by as much as 100% between one report and the next despite the fact that they appear to use similar data for the period 2000–2005. Growth rates over this period [41] are essentially exponential for large scale solar power (including both PV and CSP) with an initial capacity in 2000 of 0.2 GW rising to 0.54 GW in 2005. When this rate of growth is extrapolated to 2030, a figure for installed solar capacity, in large scale enterprises, of about 70 GW is indicated. Of this 70 GW no more than 60 GW will be accessible by the consumer, because of high transmission losses over larger than average distances, together with distribution losses. This is just a tiny fraction, namely 0.4%, of predicted demand by 2030.

3.7 Geo-thermal Power

Where the Power Comes From

Although geothermal energy is classed, in international energy tables, as a 'new renewable', it is not really a new energy source at all. Hot springs for bathing and washing clothes have been used by people in many parts of the world since the dawn of history [42]. Modern geothermal production wells can gather large amounts of power from the ground by going deep. They are commonly over

2 km deep, but at present rarely much over 3 km. With an average thermal gradient of 25–30°C/km, a 1 km deep well in dry rock formations would have a base temperature near 40°C in many parts of the world (assuming a mean annual temperature of 15°C) while at the foot of a 3 km well the temperature would be in the range 90–100°C. With sophisticated exploitation techniques, which make optimum use of these temperature gradients, it is estimated that 65–140 GW of electrical power could be generated, worldwide, from geothermal sources.

Exploitable geothermal systems occur in a number of geological environments [43]. They can be divided broadly into two groups, depending on whether they are related to young volcanoes and magmatic activity or to lower temperature mechanisms. High-temperature fields used for conventional electric power production (with temperatures above 150°C) are mostly confined to the former group, and we shall concentrate on this source. Until recently, geothermal fields were more commonly exploited for space heating purposes with direct transfer of thermal energy from the wells to local buildings. This application can be found in both groups. Needless to say, the temperature of geothermal reservoirs can vary from place to place, depending on the local conditions.

High-temperature fields capable of providing significant levels of generated electrical power are dependent on volcanic activity, which mainly occurs along so-called tectonic plate boundaries. According to the plate tectonics theory, the Earth's crust is divided into a few large and rigid plates which float on the hot inner mantle and move relative to each other at average rates counted in centimetres per year (the actual movements are highly erratic). The plate boundaries are characterised by intense faulting and seismic activity, and in many cases volcanic activity. Geothermal fields are very common on plate boundaries, as the crust is highly fractured and thus permeable, and sources of heat are readily available. While most of the plate boundaries are beneath the sea, making exploitation difficult, accessible fields exist where volcanic activity has been intensive enough to build islands and also where active plate boundaries transect continents. High-temperature geothermal fields are scattered quite regularly along the boundaries. A spectacular example of this is the 'ring of fire' that borders the Pacific Ocean (the Pacific Plate). Intense volcanism and geothermal activity associated with this fault ring is to be found in Alaska, California, Mexico, Central America, the Andes mountain range, New Zealand, Indonesia, Philippines, Japan, Kamchatka, and the Aleutian Islands. Other examples are Iceland, which is the largest island on the Mid-Atlantic boundary of the North American and Eurasian plates, and the East African Rift Valley with impressive volcanoes and geothermal resources in, for example, Djibouti, Ethiopia and Kenya.

A source of geothermal energy that is not related to the heat at the Earth's core has recently been uncovered in Switzerland and Australia [44]. In South Australia oil and gas companies prospecting in the deserts there have uncovered massive sources of heat just 4 km below the surface. This heat resides in granite strata and is generated by the natural radioactivity in the rock. The heat is trapped there by the sedimentary blanket, which extends for 4 km up to the surface. However, the exploitative potential of such sources remains to be assessed.

How the Power Is Extracted

Electricity generation stations employing geothermal techniques comprise relatively conventional steam turbines, as used in coal fired power stations, and these act as prime movers for synchronous generators. The basic process involves pumping high pressure water down a borehole in the rock into the heat zone some two or three kilometres below the surface. The water travels through fractures in the rock, capturing the heat of the rock, raising its temperature to about 150°C, until it is forced out of a second borehole as very hot water, which becomes steam as it reaches the surface. The energy in the steam is converted into electricity using a steam turbine. In lower temperature wells a secondary fluid, usually organic, with a low boiling point and high vapour pressure, is used and conversion to electricity occurs in a, so called, binary power plant (Fig. 3.9). In the pressurised water system the exhaust steam from the turbine passes to a condenser and cooling tower, and the cooled water is injected back into the ground to repeat the heating and cooling cycle. Conversion efficiencies are comparable with those of coal-fired power stations and power outputs to the grid from a single power station are typically in the range 20–50 MW.

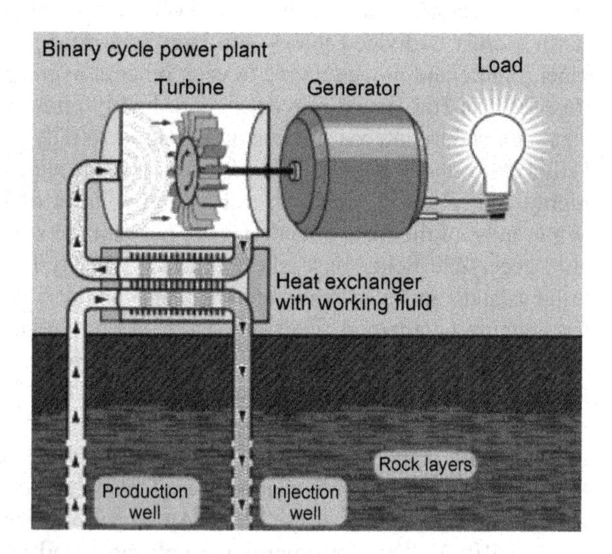

Fig. 3.9 Geothermal power extraction employing the binary operating principle

Potential as a Source of 'Green' Energy

In global terms the contribution of geothermal sources to electrical power generation is quite small at just over 9 GW in 2005. Where local conditions are favourable it is clear that electricity generated from geothermal sources is an attractive option, and capacity could, with proper encouragement, be increased fifteen-fold.

However, the evidence emanating from authoritative reports [45] on renewable energy, is that few plans are in place in 2008 to significantly increase electricity generation capacity by this means, and consequently we can rule it out as a significant contributor to global power requirements by 2030.

3.8 The End of an Illusion

One does not have to delve too deeply into the seemingly vast output of the green-wash industry, to find overly optimistic statements, bordering on 'flights of fancy', about the potential of renewables to satisfy all of mankind's energy needs. On the basis of the claims of advertising copy alone, the switch from our fossil fuel past, to a renewable power future, could happen painlessly, and now, apparently!

In both its written and electronic form, and in the guise of consultancy reports, magazine articles, press comment, energy industry websites, green websites, or other material, statements such as 'our energy requirements can be satisfied by covering just 1% of the worlds deserts with solar power collectors' are not uncommon. For example in a UK quality Sunday newspaper, *The Observer* (2/12/07), in an article on 'How Africa's desert sun can bring Europe power' the following super optimistic statement is made: 'Scientists estimate that sunlight could provide 10,000 times the amount of energy needed to fulfil mankind's current energy needs'. It at least qualifies this by saying: 'Transforming that solar radiation into a form to be exploited by humanity is difficult however'.

So how much power can renewables really provide to a rapidly growing population of increasingly energy-hungry human beings? In previous sections of this chapter, we have looked at each of the primary renewable resources in turn to assess their potential contribution by 2030 and beyond. In so doing it has been possible to show, by first assessing the magnitude of the resource as predicted by fundamental physics, followed by efficiency calculations on the collected power as it is subsequently processed through various stages of electricity production – turbines, generators, up-conversion transformers, transmission over the grid, down-conversion transformers and distribution to consumers – that, despite the 'hype', the power available to users is by no means limitless. 'Firm' estimates for ultimate electrical power levels, which can be extracted from *realistically* accessible renewable sources, are summarised in Table 3.1 (see the column labelled (2050+)). These estimates are of an accuracy that an engineer would describe as being of 'ball-park' reliability, since they are based mainly on engineering evaluations of the science and technology, but with some geographical and geological guesstimates thrown in.

The following observations are apposite:

Hydro: Hydro-electric schemes represent a mature renewable energy resource, and in the Western industrialised nations most of the viable sites for reservoirs and dams have been commandeered. Growth is occurring mainly in the new industrial nations of Asia, in particular China and India. It is difficult to assess to what extent

Table 3.1 Installed renewable power at global level: by 2030 and long term

Resource	Installed power at the point of consumption by 2030 (TW)	Available power at the point of consumption (TW) (2050+)
Hydro	0.8400	~ 2.000
Wind	0.4300	7.500
Wave	0.0003	0.022
Tidal	0.0001	0.200
Solar	0.0600	4.500
Geothermal	0.0090	0.140
Nuclear	0.8000	1.800
Total	2.1400	~16.200
Fraction [5] of (projected demand)	8.9% (~24 TW)	54.0% (~30 TW)

mankind will exploit sites that in the past might have been considered to be too difficult, too controversial, and too expensive, once fossil fuel derived energy is in very short supply. My guess is that about 2 TW, almost three times the 2008 level, is the best that could be achieved in the long term. This is of the order of 13.4% of the power (15 TW) currently consumed by mankind.

Wind: Given the extent to which land is already being commandeered by the human population, it is estimated (Sect. 3.3) that using currently available technology, an area of land and shore, about equal to the land area of Mexico, but spread across the globe, could possibly be identified for coverage by wind farms, if the desire for energy becomes sufficiently desperate. This results in the figure of 7.5 TW of power to the consumer from wind, after all loss mechanisms have been factored into the calculation. It is difficult to see how more could be extracted from wind in the long term, if we assume that the human population will continue to grow, and will require to maintain, at least at present levels, other forms of land usage. The figure is equivalent to 50% of current (2008) demand – a very significant contribution from wind – but it is predicated upon solving the variability issues.

Wave: Wave power will contribute only a tiny fraction (0.14%) of man's energy needs even in the very long term. Despite the fact that the power in the waves is vast, little is available for exploitation, unless we learn to extract it in the deep ocean. With current technology we are limited to shore based, or close to shore, collection schemes. In addition global coastlines that offer good waves in sites that are not unfeasibly hostile are estimated to be no more than 5000 km in extent. Even if all of this coastline were optimally employed as wave farms, the most that we can possibly harvest is about 22 GW.

Tidal: The gravitational physics governing tidal movements indicate that the potential energy built into the 'pull' of the sun and moon on the seas and oceans of the globe, while large (equivalent to about 130 Aswan dams), is quite limited by comparison with hydro, wind and solar resources. Like wave activity it is also very difficult to access. Suitable sites for barrage and tidal stream methods of tapping into the tides are scarce, and calculations suggest that, at best, 0.2 TW of electrical power could be extracted for tidal resources. This is 1.3% of current demand.

Solar: Using basic geometry and the known radiant power of the sun it is possible to establish a figure for the radiant power density striking the Earth's disc. This quantity is termed the solar constant and is currently estimated to be 1367 W/m^2. From this the solar power density at the Earth's surface can be computed and is generally quoted as having a mean value of 170 W/m^2. Used unwisely this statistic can generate hugely over-optimistic estimates of exploitable solar power levels. This is because, when delivery to the consumer is the criterion, and conversion, generation, transforming, transmission and distribution inefficiencies are factored into calculations, a figure of 4.5 W/m^2 is obtained for the watts per square metre of land that can be extracted from solar radiation with currently available technology. The most effective locations for solar farms are hot, arid deserts, but even these locations have other uses and are not devoid of ecological importance. Land area available for massive solar farms is not 'unlimited' and reasoned deliberation suggests that an upper limit of 4.5 TW of electrical power is available from solar sources. In the long term, therefore, about 15% of global demand (30 TW [5]) could be met from solar power stations and other solar gathering activities.

Geothermal: As with wave and tidal power, geothermal power represents a useful but small resource in global terms. Reliable estimates suggest that output from this resource, with current levels of technology, could over time possibly reach 15 times the power being delivered in 2005, which gives a ball-park prediction for geothermal power of 140 GW. Consequently, geothermal sources could potentially add 0.6% of demand to the renewables 'mix' in the long term.

Nuclear: Figures for nuclear power generation have been included in the table for completeness, although nuclear fission is not strictly renewable. The issue of nuclear power is considered in more detail in Chap. 4, where it is deemed to provide a valuable contribution to base load for a global electricity supply system.

On the basis that an unforeseen technological break-through in extracting electrical power from renewable resources, such as nuclear fusion, is not on the horizon, and that present methods will not advance much beyond current levels of sophistication, the engineering evidence strongly suggests that electrical power generated from all renewable sources, backed up by nuclear power, will, in the long term (beyond 2070), probably plateau at a level that equates to about 50% of a potential demand of ~30 TW beyond 2050, if BAU were foolishly pursued this far into the century. It is presumed that human population will plateau at 10.5 billion towards the end of the century, and that mankind continues to be in thrall to an energy profligate, consumption driven, global economic system. To improve on the 16 TW figure (Table 3.1) would either take a step-function change in technological expertise and engineering prowess, particularly with regard to operating in hostile marine environments, or an unlikely acceptance by human societies of a visual pollution and environmental degradation levels, associated with covering vast areas of land and sea with wind, wave and solar farms, far beyond today's acceptable boundaries. In my long experience, major technical advances generally take about 20–30 years to move from an idea to full practical implementation. Consequently, if it turns out that human beings have not been smart enough to give up their addiction to the energy guzzling luxuries and trappings that fossil fuels obviously provide, then the above figure implies that once the coal mines, oil wells, and gas wells are exhausted, mankind will be forced to adjust to a severe drop in power supply. Of course, on this scenario they will, in addition, have to exist in a much degraded biosphere and with all that that may mean!

In the shorter term, the available evidence of progress for the uncoordinated market led, essentially BAU approach to the transition from a fossil fuel driven economy to one based on renewable resources, which is currently being pursued, is not encouraging when viewed from an engineering perspective. The reality, as we have seen in this chapter, is that by 2030, the change-over process will have hardly advanced at all. As the middle column in Table 3.1 shows, very little of mankind's power requirements will be met from renewables by 2030 if we continue as we are doing. The conclusion has to be that in twenty or so years, electrical power from sustainable resources will total just 9% of likely demand, which by then is estimated to be about 24 TW [5]. In other words we will barely have reached a level that would at least see the total replacement of the fossil fuels used for present levels of electricity generation, with renewable resources. (In the industrialised nations electricity generation represents about 10% of total energy usage.) It seems clear that the currently popular, technology led, incoherent market driven approach to countering the global warming threat is going to fall far short of its supposed goal; namely, that by 2030 the predicted 2°C rise in mean global temperature above pre-industrial levels, should be averted. The science community, as we have observed in Chap. 1, looks upon 2°C as a critical 'tipping point', and it is a signpost that mankind would be incredibly foolish to ignore.

So is there a solution, given that the accepted scientific wisdom is calling for the threat to be addressed, and given that the market driven techno-fixes are likely to be 'found wanting'? This issue will be considered further in Chap. 5, but before that can be done, it is necessary to address the problem of energy storage, which will be of critical importance in a world reliant on intermittent renewables.

Chapter 4
Intermittency Buffers

Engineering is the science of economy, of conserving the energy, kinetic and potential, provided and stored up by nature for the use of man. It is the business of engineering to utilize this energy to the best advantage, so that there may be the least possible waste.

William A. Smith

It doesn't matter whether you can or cannot achieve high temperature superconductivity or fuel cells, they will always be on the list because if you could achieve them they would be extremely valuable.

Martin Fleischmann

4.1 Energy Storage

Products that use tiny amounts of electrical power, supplied from energy stored in batteries, such as radios, hand-held phones, laptop computers, watches, toys, etc., are commonplace, but as consumers and users will know well, this is a very expensive way to energise electrical gadgets. In fact cost effectively harnessing and storing electrical power remains a major challenge to science, particularly when very large quantities of electrical energy are involved [1]. Current generation and transmission policy means that to all intents and purposes electricity has to be used while it is being generated. When the power station generators cease to spin, for any reason, the grid wires linking them to consumers become 'dead'. In an electricity system based entirely on renewable resources this could happen frequently unless back-up power from alternative sources or from massive storage systems are available. In general, as we have seen in Chap. 3, power or energy flow from a renewable resource is not constantly available, but depends on weather conditions, or time of day, or season. Furthermore energy demand by human societies is also by no means invariant. It depends on the same phenomena but largely in reverse. So there needs to be a mediating technology between the source and the consumer. This technology is energy storage which, in one way or another, actually plays a role in all natural and man-made processes.

Storing energy in any form other than solid, liquid or gaseous fossil fuels, is a very expensive undertaking. This is because of the high capital costs of building massive storage facilities for the alternatives, which generally store energy in much lower densities. In fact a significant consequence of moving towards renewables will be that consumers will increasingly become aware that they are paying for the capacity of the particular energy storage and supply system and not for the energy itself. Their bills will increasingly become related to the size of the system built to serve their demands, rather than being related to the amount of energy they consume. In general, vast storage facilities are likely to be an integral part of power supply systems geared to exploiting renewable resources. These Massive Energy Storage (MES) systems are the critical technology needed by a renewable power generation system if it is to become a major source of readily accessible base load power, and hence eventually replace fossil/nuclear power plants. For system stability and load levelling, stored energy banks capable of releasing many megawatts of power quickly, and of providing this power over many hours, are needed to convert the intermittent and fluctuating renewable power, into electricity on demand. Without sufficient MES accessible at all times, solar/wind/wave power cannot serve as a stable base load supply; it can only piggyback onto base load fossil/nuclear generators as a small incremental supplier. That there is a need for MES is self evidently largely absent from public discussion of strategies for the development of renewable power. The prevailing public view, where it exists, is that renewable power can probably replace fossil/nuclear generating stations, if enough wind farms, wave farms, and solar generators are built. All attention, on support for research and development, is shortsightedly focused on improving the cost and performance of wind/wave/solar electricity generators. MES is not even recognised as a top priority critical technology deserving equally sustained attention and support if a 'leap' to renewables is to have any chance of being successful.

In this chapter we will consider and assess the practicability of several large scale storage systems, from pump storage commonly associated with hydroelectric schemes, through compressed air, flywheels, thermal storage, batteries, hydrogen, capacitors and superconducting magnets.

4.2 Pump Storage

Storage Principle

Pump storage is in essence a system for enhancing the operation of hydro-electric power plants, by assisting nature in refilling the reservoir. In particular, it is applicable to those schemes that operate with more than one reservoir. In these circumstances, it becomes possible, at periods when demand for power from the grid is low, to use excess generation capacity to move water from a lower reservoir to a higher one, for future use when demand is high. Water is then released back into the lower reservoir through a turbine, generating electricity in exactly the same way as for a conventional hydro-electric power station. A system of this description was first used in Italy and Switzerland in the 1890s. Now, there is a large number of pumped storage systems in operation worldwide producing over 90 GW of power to the grid. This is about 3% of current global electrical generation capacity. More specifically in Europe, in 1999, a total of 188 GW of hydropower capability was in existence, with 32 GW of it emanating from pumped storage, mainly in Scandinavia. At that time this represented 5.5% of total European electrical capacity.

Pumped storage hydro-electric systems, like their conventional counterparts, use the potential energy possessed by water when it is raised against the force of gravity, the primary difference being that mechanical intervention is employed to elevate the water. As we have already noted in Sect. 3.2, the potential energy density in stored water is very low and therefore it requires either a very large body of water or a large variation in height to achieve substantial storage capacity. For example, in the Cruachan system in the west of Scotland, water is pumped from Loch Awe to the upper reservoir below Ben Cruachan, 360 m above, during periods of low consumer demand (such as at night). A 316 m long dam forms the upper reservoir, which contains about $13 \times 10^6\,m^3$ of water. Additionally the upper reservoir collects substantial amounts of rainwater. Tunnels have been built through Ben Cruachan to catch rain coming from all sides of the mountain. Around 10% of the energy from the station is generated from rainwater, the rest is from the water pumped up from Loch Awe. The above statistics suggest that the potential energy contained in the upper reservoir is of the order of 46,000 GJ. Operating at full capacity, the 440 MW Cruachan hydro-electric power plant would deplete the upper reservoir in 22 hours, although the power station is required to keep a 12 hour emergency supply. Replenishing the reservoir through pumping, with the generators acting as motors and the turbines as pumps, involves the raising of $6 \times 10^6\,m^3$ of water from Loch Awe, 360 m below. This represents an energy input of 21,000 GJ. Assuming 400 MW of power is available in pumping mode, it will take about 14.6 hours to do this. However, since system efficiency is at best about 75%, we will actually require 530 MW to be supplied from the grid to complete the task. In the case of the Cruachan system, this means essentially from the nearby Hunterston nuclear power station. Of course, this is a worst case scenario. In the rain sodden west of Scotland, water hardly ever stops streaming off the hills and bens, and reservoir replenishment is also occurring naturally. If we follow through the full pumping/power-delivery cycle, from the power required to replenish the stored energy (75% efficient), to the depletion of this energy in supplying consumers (70% efficient), almost 50% of the power generated disappears in electrical system losses. Even so, the economics of large scale electricity generation has determined that rapid and ready access to hydro-electric power justifies this seemingly high level of wastage.

Technology Required

Reversible turbine/generator assemblies acting as pump and motor first became available in the 1930s. These turbines can operate as both turbine-generators and in reverse as electric motor driven pumps. Of course, considerable advances have been made over the last half century, and the latest developments in large scale turbine technology are variable speed machines, which promise much greater efficiency. Importantly, these machines facilitate electricity generation in synchronisation with the grid frequency of 50/60 Hz, yet can operate asynchronously (independent of the network frequency) as motor-pumps. They are usually of a design described as a Francis turbine, which functions using the reaction principle of operation. This makes it amenable to operation as both a pump and a turbine. It is not unlike the Kaplin turbine described in Sect. 3.2.

Potential for Providing Intermittency Correction

Like conventional hydro-electric schemes, pumped storage plants are characterised by long construction times and high capital expenditure. Nevertheless, pumped storage is the most widespread, large scale, energy storage system currently in use on power networks. Its main applications are for energy management, smoothing variable demand and provision of reserve. But in addition, these systems help to control electrical network frequency. Thermal and nuclear plants are poor at responding to sudden changes in electrical demand, potentially causing frequency and voltage instability on the grid. Pumped storage plants, like other hydro-electric plants, can respond to load changes within seconds thus helping to reduce the problems caused by short term variations in demand. Pump storage systems can be found which range in scale from the compact to the very large with discharge times varying from several hours to a few days. The critical factor influencing the slow development of such schemes is high capital costs and the dwindling existence of sites providing appropriate geography and geology.

Although 50% of generated power can be lost in pumping and power delivery, pump storage is considered to be an acceptable use of electrical power because it flattens out load variations on the grid, permitting thermal power stations such as coal-fired plants and nuclear power plants that provide base-load electricity to continue operating at peak efficiency, while reducing the need for 'peaking' power plants that use costly fuels. Today, however, with current knowledge of the ecological harm being caused by the burning of fossil fuels, the economic argument for pumping using thermal plants becomes much more difficult to sustain. In future, when all of our energy is derived from renewables, pumped storage, and other methods of energy storage will become the primary source of electrical power continuously being 'topped up' by renewable power stations, thus negating the fluctuating output of these intrinsically intermittent power sources. The storage system will absorb load at times of high output and low demand, while providing additional peak capacity. A high percentage of renewable power, without a parallel MES system, could result in electricity prices dropping to close to zero, or even occasionally going negative, as happened in Ontario in early September, 2006. More power was being generated (some of it from wind) than there was load available to absorb it. At present this sort of event is rarely due to wind alone, but as the proportion of our electrical power from renewables grows, the frequency of such occurrences is likely to increase, unless a mediating electricity storage system exists to absorb excesses.

The question is then – how can pump storage advance the transition to renewables? It could be said that this is possibly happening already, since this technology is clearly extending the reach of hydro-electric power. However, in relation to other renewable resources pump storage is actually a rather inflexible technology since it is very dependent on geology. All of the best sites for its implementation have already been commandeered. Nevertheless pump storage formats, which would in earlier times have been viewed as impractical and uneconomic, are now being looked at with renewed interest. One of these is underground pumped storage, using flooded mine shafts or other underground cavities. The arrangement has been shown to be technically possible and is being pursued quite actively in several parts of the world. The open sea can also be used as the lower reservoir in a pumped hydro-system. The first seawater pumped hydro plant, with a capacity of 30 MW, was built in Japan, at Yanbaru, in 1999, and other schemes are being planned. Additionally pump storage has been proposed as one possible means of balancing power fluctuations from very large scale photovoltaic and CSP generation [2]. On a European scale it is suggested that through the agency of high voltage direct current (HVDC) transmission, it would be economic to recharge pump storage facilities in Scandinavia from excess solar power generated in North Africa.

4.3 Compressed Air

Storage Principle

Air is an elastic medium, and when compressed it stores potential energy. Released air, if expanded in a controlled manner, can be used to power a gas turbine, and consequently the use of compressed air energy storage (CAES) for power utilities has been under consideration for a very long time. Such a system, employing an underground cavern to store the air was patented by Stal Laval in 1949. Since then two operational plants have been completed and commissioned – one in Germany (Huntorf CAES) [3, 4] and another much more recently in the USA (McIntosh CAES) [5].

The principle of energy storage in compressed air is really quite simple. Anyone who has done some school chemistry will be familiar with Boyle's Law of gases [6], which states that for a gas of constant mass the product of its volume and its pressure is proportional to its temperature. Consequently for a gas in a chamber, which is being compressed by the movement of a piston, the gas pressure exerts a force on the piston. This force (newtons) is equal to pressure (pascals) times the area of the piston (m^2). In overcoming this force to move the piston, work (in joules) must be done, which, for small movements, is equal to the force times distance moved. By Boyle's law the change in volume and the increase in pressure will produce an increase (or decrease) in temperature, and the storage of some heat in the gas. The first law of thermodynamics then dictates, by conservation of energy, that the applied work on the piston must equate to an increase in heat stored plus the stored elastic energy in the compressed gas. Usually the change in temperature can be assumed to be small (isothermal operation) in which case the stored energy in the gas can be readily calculated [7]. For example, if 1000 m^3 of air at 2.03×10^5 Pa is compressed at constant temperature so that its volume is reduced by 60% then the elastic energy stored in the gas will be 0.186 GJ or 0.052 MW-h. Larger cavity or chamber volumes will produce proportionally larger stored energy levels. Very large storage volumes of the order of 500,000 m^3 with air at pressures in the range 7–8 MPa have been proposed to procure energy storage levels in excess of 500 MW-h. However, the only practical way of storing volumes of this magnitude is to use impermeable underground caverns at depths of 700–800 m.

Technology Required

An electricity supply plant operating with a compressed air facility as back-up would function as follows [8]. A compressor, a small version of which is to be found in every fridge/freezer, draws power from the electricity supply system, during a demand trough. Air at atmospheric pressure passing through the intake aperture is then compressed to a high pressure before being forced into the deep underground storage cavern. At times of peak demand the compressed air is piped from the cavern and energy is released when it is mixed with fuel and ignited in a combustor. In a renewable power system, of course, this fuel would have to be bio-generated. The resulting high energy gases are then directed over the blades of the turbine(s), spinning the turbine, and mechanically powering the electricity generator(s). Finally, the gases are passed through a nozzle, generating additional thrust by accelerating the hot exhaust gases as a result of their rapid expansion back to atmospheric pressure through a second turbine. The efficiency of this storage process involving, as it does, air compression by pumping, which requires power expenditure, followed by energy release through a turbine, is of the order of 20–40%. But, in much the same way as for other storage techniques, if pumping is done during periods of low demand when electricity is 'cheap', the process becomes economically justifiable. This is even more valid when renewable resources are being employed.

The volume of the compressed air reservoir is obviously determined by the power station supply/demand cycle. The volume will tend to be sized to, for example, ensure that the turbine(s) can run for an hour (say) at full load, while the compressor will be designed to replenish the reservoir in the average duration of low demand – typically about 4–5 hours. So the compressor is sized for only a quarter of the turbine throughput, which results in a charging ratio of 1:4. Charging ratios from 1:1 to 1:4 are not difficult to accommodate with modern equipment and therefore a reasonable degree of operational flexibility is available to fit in with the geological conditions of any given site possessing a suitable underground cavern.

Undeniably CAES is severely limited by geology. Airtight caverns with volumes in excess of 100,000 m^3 are hard to find or to form deep underground. Three types of cavern are generally favoured. These are salt caverns, aquifers, and hard rock cavities [9]. The forming of large shaped cavities in natural salt deposits by 'solution mining' has been possible for some considerable time. It was first developed for storing natural gas and also waste materials in order to seal in noxious gases. There is, in fact, growing experience in Europe and in the USA, of salt caverns being used to store gas, oil and other substances. The method of excavation, namely solution mining, represents a relatively cheap method of creating very large cathedral like volumes underground. Furthermore, for gas storage such caverns are practically 'leak tight'. For example, it is estimated that the two salt cavities at Huntorf leak no more than 0.001% of the volume of the air in each cavity, per day. The technology of solution mining is based on fresh water dissolving the salt and becoming saturated with it. The water is forced into the salt deposit through a cylindrical pipe, centred in a lined bore hole of slightly larger diameter, drilled from the surface. The saturated water solution, or brine, is forced to the surface through the annular space between the pipe and the bore hole. Various techniques are used to control the shape of the cavity, which should ideally be in the form of a vertical cylinder whose height is about six times its diameter, to minimise any chance of collapse. An awful lot of brine is produced, which can result in a disposal headache if one is playing by ecologically friendly rules!

Salt layers, at suitable depths, with sufficient thickness and in locations where storage plant is required, are not uncommon in Europe and in the USA but this is not the case in the rest of the world. Japan is particularly poorly served, apparently [9]. However, underground cavities for storing compressed air can be created in other ways. One of these is based on the use of aquifers, i.e., large natural caverns containing water. Provided the cavern has a domed impermeable cap rock it will be suitable for gas storage [10]. The basic requirement for successful storage is the formation of an 'air-pocket' between the subterranean lake and the roof of the aquifer. If this is not available naturally, multiple wells may be required, to first form the air pocket, and second to maintain it. Leakage statistics for this method of storage are less favourable than for salt caverns.

The third possible approach to the forming of underground cavities for compressed gas storage is straightforward mining of hard rock. It is potentially the most expensive of the three options, insofar as the mining can be difficult and time consuming and the disposal of very large amounts of debris can also involve costly processes. Nevertheless, the high cost is off-set by the flexibility afforded by a purpose built cavity. For example, operating the storage system under constant pressure, which would involve partially filling the cavity with water linked to a surface reservoir [7], enables much more efficient turbine operation from the compressed air source. Leakage is likely to be a problem with this option, since leaktight rock strata are hard to find. Partial filling with water as suggested above will help, as will cavern lining but this adds significantly to cost. As yet, no storage facilities of this type have been constructed.

Potential for Providing Intermittency Correction

In renewable energy terms a major disadvantage of CAES, apart from the cathedral-like underground caverns, is the requirement to burn a fossil fuel to expand the air powerfully through the turbine/generator set. However, it is possible that a solution could lie in the use of synthetic fuels such as methanol, ethanol or even hydrogen, although development work seems to be required to establish these possibilities. Nevertheless, CAES, if the appropriate geological circumstances are present, is sufficiently well developed, as the Huntorf and McIntosh facilities confirm, to be a serious player in the storage mix required by an electrical power industry dependent wholly on renewable resources.

4.4 Flywheels

Storage Principle

It has been known for centuries that it is possible to store energy in kinetic form, for short periods of time, in the motion of a heavy mass. The movement commonly used to do this is that of a spinning disc or flywheel. Rather surprisingly, since it is not obvious that a flywheel could be made to spin for hours, or even days, it is a storage method that is now being re-evaluated in the light of advances in material and bearing technology, for roles more commonly associated with batteries. Composite materials reinforced with carbon and glass fibre, and new ‘hard’ magnetic materials, permit higher spin speeds, on ‘frictionless’ bearings, with lighter flywheels, and this has resulted in a rekindling of interest in applying an old technology to a new and pressing storage problem [7, 11]. Incidentally, this is a not uncommon engineering process, and is arguably one of the primary mechanisms underpinning many advances in technology.

The flywheel employs what is termed an inertial energy storage method where the energy is stored in the mass of material rotating about its axis. There are plenty of historical examples. In ancient potteries, the potter's rotating heavy table (essentially a flywheel) was kept turning at fairly constant speed by an occasional and judicious kick from the operator, at a protruding floor level rim to the table. The energy of the kick was sufficient to maintain rotation. The rotating mass of the table stores the short energy impulse and if the mass is heavy enough, and if the friction is low, the table will spin at a steady and reasonably constant speed. During the steam age, of course, flywheels were very common, being widely applied to reciprocating steam engines in order to smooth the uneven power delivery from the piston. Steam traction engines with external brass and steel flywheels were once a quite familiar site, during the last century, on the roadways and byways of the industrialising world. In the 1950s flywheel-powered buses, known as gyrobuses, were introduced into service in Yverdon, Switzerland, while flywheel systems have also been used recently in small experimental electric locomotives for shunting or wagon switching operations. Some large electric locomotives, e.g., the BR Class 70 locomotives in the UK, are fitted with flywheel boosters to carry them over occasional quite large gaps in the 'third' power rail of the electric drive system. More recently, flywheels such as those incorporated into the 133 kWh pack developed at the University of Texas at Austin have been demonstrated to be sufficiently advanced, to power a train from a standing start up to full cruising speed [12]. It is clear that flywheel energy storage levels are steadily being extended as a result of well established and quite active research and development in this branch of engineering.

The above examples represent traditional flywheels in both short term storage and smoothing applications. The re-examination of flywheels for more extended storage roles is a quite recent development. Today it is becoming realistic to apply them to the storage of energy over long time intervals. For protracted storage of this kind, flywheel design has had to advance on several major fronts; namely flywheel shaping, material composition, low loss bearings, and evacuated containment vessels [13]. Each of these crucial elements of flywheel development will be assessed briefly below, with the aim of appraising the viability of the flywheel as a storage medium of relevance to the delivery of electrical power from renewables.

Technology Required

We have already shown, in Chap. 2 in relation to pendulum motion, that the kinetic energy in a weight moving with a linear velocity is given by half the mass multiplied by the velocity squared. It turns out that the stored kinetic energy in a disc or cylinder, rotating about its axis, can be approximated by the same formula [14] if the linear velocity is replaced by the tangential velocity of the disc's rim. If the revolution rate is known, usually in revolutions per minute (rpm), then the tangential velocity of the rim for a disc of radius R is given by 0.033 π times rpm times R, the result being in m/s. A conservatively designed flywheel formed from a 5 m diameter and 1.5 m thick disc spinning at say 250 rpm will have a rim velocity of 64.8 m/s. The disc has a volume of 29.45 m^3, which means that its mass is 235,600 kg given that steel has a density of 8000 kg/m^3. Therefore the energy stored in the flywheel is of the order of 495 MJ, which equates to 0.14 MW for an hour (0.14 MW-h). 500 MJ of energy stored in a flywheel has been demonstrated [7] by the EZ3 short pulse generator at the Max-Planck Institut fur Plasmaphysik at Garching, in Germany, which delivers 150 MW of electrical power in an 8 s discharge time. In theory, more energy could be stored by making the flywheel bigger or by spinning it faster. However, there is a limit to what is possible, set by the maximum tensile stress that the material forming the flywheel can sustain – in the case of steel about 900 MN/m. Nevertheless, stored energy could easily be increased by a factor of about 50 for the above flywheel by shaping it to distribute the weight, thus ensuring that the tensile strength limits for the steel are not exceeded. The tensile stress is associated with the high spin rates and it can be reduced by distributing the centrifugal forces more evenly through the volume of the flywheel, usually by thickening the disc near the axis and reducing its thickness near the rim. Such a flywheel could store over 5 MW-h of energy and, importantly, this energy can be extracted rapidly and efficiently. Actual delivered energy depends on the speed range of the flywheel. It obviously cannot deliver its rated power if it is rotating too slowly. Typically, a flywheel will deliver ~90% of its stored energy to the electric load, over a speed range of the order of 3:1.

An alternative approach to the storage problem, which is being investigated strenuously, is to store high levels of energy in low weight, high speed flywheels, by employing advanced composite materials to withstand the high stress levels (Fig. 4.1). It is predicated on the use of a wheel design comprising several radially spaced concentric rings [7]. The rings are hoop-wound from Kevlar-fibre/epoxy, and carbon-fibre/epoxy layers, which have been compression-stressed in the radial direction. It thus becomes possible to store energy in large amounts in a relatively

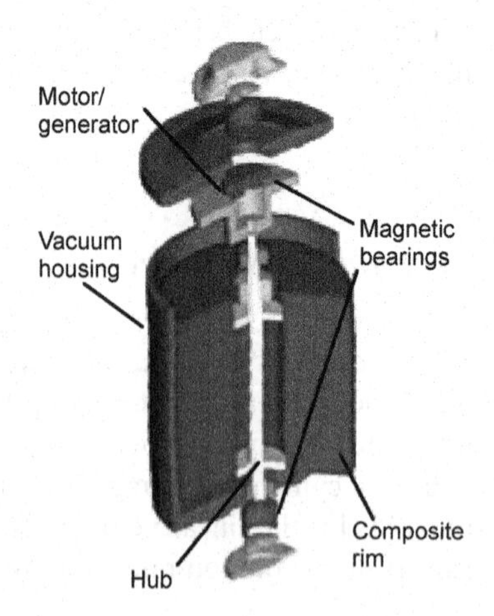

Fig. 4.1 Shematic of flywheel storage system showing generator and magnetic earings (www.electricitystorage.org/photo_flywheels1.htm)

light flywheel. To store 1 MW-h of energy would require a spin speed of about 3000 rpm in a wheel with an outer diameter of 5 m and an axial length of about 5 m. Such a wheel would have a total mass of 130,000 kg (about 140 tons), giving a storage density of ~28 kJ/kg. For electrical storage applications, the flywheel is typically housed in an evacuated chamber and connected through a magnetic clutch to a motor/generator set out-with the chamber. In turn, through the agency of some power electronics to stabilise frequency, the generator interacts with the local or national grid. Potentially, tens of megawatts can be stored for minutes or hours using a flywheel farm approach. For example, fifty vertically mounted 1 MW-h wheels, in 20 ft deep pits, could store 0.18 TJ efficiently in a relatively small footprint of about 6000 m^2, and with no more visual impact than a low level warehouse.

Arguably, the evolution of the magnetically levitated bearing has been most influential in engendering intense new interest in flywheel energy storage, particularly in relation to moderating power supply variability inherent in electricity generation from renewables. Magnetic levitation takes advantage of the Lorentz force (see Chap. 2) which occurs when a permanent magnet (incorporated into the flywheel shaft) is in close proximity to current-carrying coils (built into the stator). The Japanese Maglev trains that have created so much interest in recent years, use the same force. In conventional mechanical bearings, friction is directly proportional to speed, and at the kind of speeds proposed for storage flywheels, far too much energy would be lost to friction. In idling storage-mode, the flywheel would quickly slow down, uselessly losing energy to bearing and air friction. Consequently, low loss magnetic bearings are critical to the viability of energy storage in high speed, heavy flywheels. But levitated bearings employing the Lorentz force can also incur losses associated with the currents flowing in the lifting and stabilising stator coils. This joule heating loss can be reduced by using coils formed from low temperature superconductors, but this requires very cold operation of the bearings. The expense of refrigeration has led to the early discarding of this solution. Current research into high-temperature superconductor (HTSC) bearings is more promising, indicating that this solution is potentially more efficient and could possibly lead to much longer energy storage times than has hitherto been seen. However, flywheels employing hybrid bearings are most likely to appear in early applications. In these hybrid embodiments, a conventional permanent magnet levitates the rotor, but the high temperature superconducting coils keep it stable. If the rotor tries to drift off centre, a reaction force due to a balancing magnetic flux restores it. This is known as the magnetic stiffness of the bearing. Superconducting coils are particularly effective in stabilising the floating rotor because the magnetic force between the rotor permanent magnet and the encircling coils is controllable by small adjustments of the current in each coil in quick response to signals from sensors monitoring the bearing alignment. The coils, sensors and the intervening electronics form a control system that maintains the alignment. On the other hand, HTSC bearings have historically had problems providing the lifting forces necessary to levitate these large heavy flywheel designs because coil current levels required to procure flotation are beyond the capability of present day electrical generators. Therefore, in hybrid bearings, permanent magnets provide the levitating function while HTSCs perform the stabilisation role.

As we have seen, one of the primary limits to flywheel design is the tensile strength of the material used for the rotor. Generally speaking, the stronger the disc, the faster it may be spun, and the more energy the system can store. But this storage benefit creates a significant problem; namely, if the tensile strength of a flywheel were to be exceeded the flywheel is likely to shatter dramatically, releasing all of its stored energy at once. This uncommon occurrence is usually referred to as 'flywheel explosion', since wheel fragments can attain kinetic energy levels comparable with that of a bullet. Consequently, very large flywheel systems require strong containment vessels as a safety precaution. This, of course, increases the total complexity and cost of the device. Fortunately, composite materials tend to disintegrate quickly once broken, and so instead of large chunks of high-velocity shrapnel one simply gets a containment vessel filled with red-hot powder. Still for safety reasons, it is usually recommended that modern flywheel power storage systems be installed below ground level, to block any material that might escape the containment vessel, if an 'explosion' should occur.

Residual parasitic losses such as air friction in the imperfectly evacuated containment vessel, eddy current losses in magnetic materials, and joule losses in the coils of the magnetic bearings, in addition to power losses associated with refrigeration, can all limit the efficient energy storage time for flywheels. Improvements in superconductors should help to eliminate eddy current losses in existing magnetic bearing designs, as well as raise overall operating temperatures eliminating the need for refrigeration. Even without such improvements, however, modern flywheels are potentially capable of zero-load rundown times measurable in weeks, if not months. (The 'zero load rundown time' measures how long it takes for the device to come to a standstill when it is not connected to any other devices.) Over time, the flywheel will inevitably slow down due to residual frictional losses and bearing losses, which are impossible to suppress completely. For example, a 200 ton flywheel, in the absence of technological improvements such as those described above, would lose over half of the energy stored in it, in a 24 hour period, due to bearing and other losses [7]. Despite the difficulties, flywheel storage systems are undergoing intensive development because of their potentially very high efficiencies [7] (85%) compared with many of the alternatives.

Potential for Providing Intermittency Correction

In renewable energy storage terms, interest in flywheel technology is further boosted by other key features such as minimal maintenance, long life (at least 20 years or tens of thousands of accelerating /decelerating cycles), and environmental neutrality. It is clear that modern low friction flywheels exhibit the potential to bridge the gap between short term smoothing and long term electrical storage applications with excellent cyclic and load following characteristics. The choice of using solid steel versus composite rims is largely based on system cost, weight and size. The performance trade-off is between using dense steel with low rotational rate (200 to 375 m/s tip speed) as against a much lighter but stronger composite that can achieve much higher rim velocities (600 to 1000 m/s tip speed) and hence significantly higher spin rates. While currently available models are only suitable for small scale storage, this is changing, and in time they could perhaps be employed in localised domestic and community scale roles, and with modern high speed flywheels potentially offering storage capabilities in the region of 1 MW-h, power network roles are also becoming a realistic possibility. If stored energy levels from flywheel farms can be lifted to 0.1 TJ or more, this storage method will be approaching a capacity that is of real significance to the problem of moderating and balancing electricity supply, particularly when it becomes based wholly on renewables, as it hopefully will, in the not too distant future.

4.5 Thermal Storage

Storage Principles

Energy may be stored in one of six primary mechanisms; namely potential energy (gravity, elastic), kinetic energy (dynamic), thermal energy, chemical energy (batteries), magnetic and electric fields. However, since in so much of our present day economy, energy is produced and transferred as heat, the potential for thermal energy storage (THES) merits serious examination as a facilitator for a future economy based on renewables.

Thermal energy storage generally involves storing energy by heating, melting or vaporising a material, with the energy being recoverable as heat by reversing the process [15]. Storing energy by simply raising the temperature of a substance is termed, rather curiously, sensible-heat storage. Its effectiveness depends on the specific heat (heat energy in joules per unit kilogram per degree Kelvin above absolute zero) of the substance and, if volume restrictions exist, also on its density. Storage by phase change, that is by changing a material from its solid to liquid phase, or from liquid to vapour phase, with no change in temperature, is referred to as latent-heat storage. In this case the specific heat of fusion and the specific heat of vaporisation, together with the phase change temperature, are significant parameters in determining storage capacity. Sensible and latent heat storage can occur simultaneously within the same material, as when a solid is heated (sensible) then melted (latent), and then raised further in temperature (sensible).

Storing energy in the form of heat is probably the most common and widespread of all storage techniques particularly at domestic and factory level, and it is not surprising that much has been written about it [16]. Here, however, we will concentrate only on those techniques that are applicable to electricity power stations, with the potential to provide energy storage for matching supply to consumer demand. This is a much less common activity. For this application, thermal energy storage systems are clearly most effective as adjuncts to power stations that already employ heat to generate electricity. In conventional terms this means fossil fuel and nuclear fuel burning plants, or in a renewables scenario, it means solar power or geothermal power stations. In both cases, in periods of low demand, steam, which would normally by used to drive the turbine/generator sets, is diverted into heating a fluid in suitable storage tanks. The questions then are – what are the best storage media, what are the most suitable storage arrangements and what levels of energy can be stored?

Technology Required

Storage media choices are dictated in the first instance by the two fundamental thermal storage mechanisms defined above. First, sensible-heat storage depends solely on the heat capacity of the medium, and therefore requires a large volume or mass of the storing material with as high a specific heat as possible. The requirement for large volume dictates the use of materials that are plentiful such as water, rock or iron. The specific heats for these substances are respectively, 4180 J/kg/K, 900 J/kg/K and 473 J/kg/K, so, not surprisingly, water is most commonly employed in this kind of storage. For water at 100°C or 373 K, and given that its density is ~1000 kg/m^3, it is not too difficult to determine that 1.56 TJ (0.43 MW-h) of energy can be stored in 1000 m^3 of water, that is in a tank about the size of a swimming pool of modest proportions full of boiling water. If the hot material can be contained at high temperature over time then useful capacity for power system moderation is potentially available from this source. Water stored at high temperature has the advantage that in power station usage where the turbine/generator set is powered by steam from a boiler it can be introduced directly into the steam generation cycle without interface equipment. The main disadvantage is that above 100°C it requires a pressurised containment vessel, which is costly. With bulk concentrations of rock, or iron, on the other hand, high temperatures can be stored at close to atmospheric pressure, but finding or forming suitably large volumes, in useful locations, is an obvious drawback for these media.

The second mechanism, latent-heat storage, has been investigated using a range of different materials, which have in common relatively high specific latent heats of vaporisation or fusion. The most common of these are, with specific heats in parentheses [7]: ice (fusion = 0.335 MJ/kg), paraffin (fusion = 0.17 MJ/kg), salt hydrates (fusion = 0.2 MJ/kg), water (vaporisation = 2.27 MJ/kg), lithium hydride (fusion = 4.7 MJ/kg) and lithium fluoride (fusion = 1.1 MJ/kg). It is clear that water vaporisation or condensation provides one of the most energy-rich phase changes. In this case a 1000 m^3 pressure vessel containing steam at 100°C will release a very useful 2.27 TJ of energy (0.63 MW for an hour) when condensed into water, assuming the energy can be delivered 100% efficiently. Unfortunately constructing a pressure vessel of this size is not currently feasible at an acceptable economic cost.

Some readily available inorganic salts such as fluorides have been considered as thermal storage media since they have high specific latent heats of fusion, although in some cases at very high temperatures in excess of 800°C. Such high melting temperatures are a big disadvantage since they are a cause of severe corrosion problems. Eutectic mixtures, which retain the useful specific heat property of the original fluoride but at lower temperatures, have been proposed to circumvent this difficulty. An example is lithium-magnesium-fluoride, which has a high but less corrosive melting temperature of 746°C. These salts have been extensively investigated in relation to high temperature nuclear reactor applications, and certain nitrate/nitrite mixtures have been widely used as heat transfer fluids in moderate temperature industrial storage applications. Thermal storage in salt hydrates, such as Glauber salt, is the most commonly employed medium after water. In water at about 32°C this salt dissolves, forming sulphates of sodium plus heat at the level of 0.252 MJ/kg. Because they have a much higher density than water, the storage capacity of salt hydrates is much higher per unit volume over a small temperature range, which means that they could provide a route to much more economical storage systems, given that much of the cost of thermal storage is bound up in the complexity and size of the containment vessels or ponds.

Strong interest is currently being displayed by the electrical supply industry in a storage technique based on the use of a liquid combination comprising the plentiful, and non-corrosive fluids, water and methane. This combination can be stored almost indefinitely at room temperature. On a solar power plant at a period of low demand, diverting heat through a mixture of methane and water will produce a chemical reaction that generates carbon monoxide and hydrogen. At room temperature in a separate porous storage medium these gases will not interact and they can be held in this form for a very long period of time. However, at periods of high electricity demand, if hot air from the power station is passed through the porous store, methanation occurs (i.e., the gases combine to form methane and water) and in the process significant amounts of heat (through latent heat of condensation) are generated, which can be employed to boost power station output. A great deal of research [17] is being carried out into thermal reactions of this kind, where the reactive components can be held at room temperature, and for long periods of time. Storage volumes required to trap significant amounts of energy, are similar to those of water storage systems using latent heat of condensation, but with the big advantage of ambient temperature confinement of the fluids/gases.

A key criterion in assessing the practicability of a thermal storage method is cost of containment, since very large volumes can be involved. The following options are available [7]: steel tank pressure vessels; pre-stressed cast iron vessels; pre-stressed concrete pressure vessels; underground excavated cavities, steel lined, with high temperature, high strength concrete for stress transfer between liner and rock; underground excavated cavities with free-standing steel tanks surrounded by compressed air for stress transfer to the rock; underground aquifers of water-saturated sand and gravel confined to impermeable layers. The importance of matching the storage medium to the method of confinement to maximise storage efficiency and to minimise cost, is quite clear, given the commitment involved in simply creating a suitable vessel.

Potential for Providing Intermittency Correction

The evidence is that there already exists a healthy range of possible options for thermal storage, and that with well directed research, and with significant investment, this method can provide a useful and important tool in the growth of the renewable electricity supply industries. This is particularly true for the suppliers of solar power and geothermal power, for which thermal storage techniques are particularly appropriate.

4.6 Batteries

Storage Principle

In the electrical power industry, energy storage in batteries represents a well established technology. However it is a technology that is undergoing a renaissance after a forty year developmental plateau. This has been triggered by the renewables revolution, but has been facilitated by developments in power electronics and control engineering, which means that highly sophisticated battery conditioning systems can be realised at moderate cost. Modern power electronic switching processes also make it possible for an intrinsically DC battery storage systems to be easily and efficiently connected into the AC grid system.

Batteries have several advantages over some other large scale storage systems. First, because of their outstanding power and voltage controllability, they are ideal for ensuring that the generated frequency of a power station remains stable during demand surges by providing rapidly available back-up power. Second, they are very quiet and ecologically benign. Third, battery banks with a wide range of capacities can be readily constructed from factory-assembled modules. This offers storage flexibility, which does not exist with some other techniques. Finally, because battery banks can be assembled on relatively compact sites, they can be located at or close to distribution substations, rather than at a power station. This offers the benefit of avoiding transmission losses in the grid.

In Sect. 2.4 it was observed that electrical energy is formed and stored when negative charge is separated from positive charge. To do this, work requires to be done to overcome the force of attraction between oppositely charge particles, and this work, in the absence of losses, manifests itself as potential energy, which is stored in the resultant electric field. In electrochemical energy storage (ECES) systems the work of charge separation is performed by chemical processes associated with strongly reactive materials, and three different storage systems can be identified. These are primary batteries, secondary batteries, and fuel cells. Primary and secondary batteries utilise the chemical materials that are built into them, while fuel cells have the chemicals (fuel) delivered to them from outside to do the work of charge separation. Primary batteries are not rechargeable and are not relevant to bulk storage developments. We will therefore concentrate our attention on secondary batteries. From this point on a 'battery' implies a rechargeable battery, and generally those that are sufficiently large to be relevant to electrical power systems will be our focus.

Technology Required

Batteries, and fuel cells, essentially comprise two electrodes immersed in a chemical solution (usually) termed an electrolyte, while externally the electrodes are connected to an electrical circuit. An electrolyte is any substance containing free ions (an atom or molecule having lost or gained one or more electrons relative to its normal complement) and thus behaves as an electrically conductive medium. Because they generally consist of ions in solution, electrolytes are also known as ionic solutions, but molten electrolytes and solid electrolytes are also possible. The most common manifestation is as solutions of acids, bases or salts. Electrolyte solutions are normally formed when a salt is placed into a solvent such as water and the individual components dissociate due to the thermodynamic interactions between solvent and solute molecules, in a process called solvation. For example, when table salt, NaCl, is placed in water, positively charged sodium ions and negatively charged chlorine ions are formed [18]. In general terms, an electrolyte is a material that dissolves in water to yield a solution that conducts an electric current. It may be described as concentrated if, in solution, it has a high concentration of ions, or dilute if it has a low concentration. If a high proportion of the solute (e.g., salt) dissociates to form free ions, the electrolyte is strong. On the other hand, if most of the solute does not dissociate, the electrolyte is weak.

The process of charge separation and energy accumulation in a battery can probably best be explained by reference to a class with which most people will be familiar – namely the lead–acid battery which provides starting power in road vehicles. This chemical storage format has been around a very long time [19], in electrical engineering terms, having been invented by Gaston Plante in 1859. In the lead–acid storage cell the cathode is formed from spongy lead (Pb), while the anode is also made of lead but coated with lead dioxide (PbO_2). The two electrodes are usually interleaved to expose maximum surface with alternating anode and cathode surfaces. These plates are immersed in an electrolyte, which comprises a solution of sulphuric acid (H_2SO_4) diluted with water (H_2O). In a fully charged battery the proportions are 25% acid to 75% water. Lead reacts quite strongly with sulphuric acid to form lead sulphate ($PbSO_4$) and water [20]. During the reaction, which occurs because of the dictates of the second law of thermodynamics (Sect. 2.2), free electrons are formed at the cathode and work is performed on these electrons moving them from the cathode to the anode, which becomes negatively charged. In so doing the cathode becomes deficient of electrons and hence positively charged. In a battery unconnected to an outside electrical circuit (open-circuited) the reaction will continue until the potential between each pair of plates (2.014 V) matches the chemical potentials driving the reaction. This voltage is usually referred to as the electromotive force (emf). Note that in relation to the outside circuit the cathode plates are connected to the positive terminal of the battery, while the anode plates are connected to the negative terminal. With six anodes and six cathodes (six cells) the battery will deliver 12.084 V. Furthermore, if it is connected to an electrical load, such as a car starter motor, current can be drawn until both the anode and cathode are fully coated with lead sulphate at which point the electrolyte has also become highly diluted with water. The process can be reversed and the battery recharged by passing a DC current through the battery such that electrons are made to flow from the anode to the cathode [20].

The electrical energy provided by a battery during discharge is derived from the electrochemical reactions taking place between the electrolyte and the active materials in the anode and the cathode. In the case of a lead–acid battery the reaction is between the sulphuric acid and the lead in the cathode and the lead dioxide in the anode. The greater the amount of active material the greater is the storage capacity of the battery. The electrochemical laws of Faraday [21] provide the method of calculating these amounts, and when applied to the lead-acid battery yield the result that for 1 A-h of electrical capacity, 4.46 g of lead dioxide and 3.87 g of lead is required [17]. In practice, from three to five times these theoretical amounts is needed, depending on the type of cell and the thickness and number of plates. Given that 1 A-h from a 12 V battery represents 12 W-h or 43.2 kJ, then we can conclude that a lead–acid battery has a storage capacity of the order of 0.61 MJ/kg. This is similar in level to thermal storage based on phase change techniques (see Sect. 4.4), but is more than twenty times greater than is currently available from flywheel storage (28 kJ/kg), and greatly exceeds the per kilogram values associated with hydro-electric pump storage. On the other hand, petrol has an energy storage capacity of 47 MJ/kg, while hydrogen provides 143 MJ/kg. It is hardly surprising, therefore, that mankind has largely ignored renewable energy sources in favour of fossil fuels.

While weight for weight, or volume for volume, batteries tend to be the most compact of electrical energy storage media, transference of energy into and out of a battery generates rather significant levels of power loss, which can represent a major problem for some storage applications. If we consider, for the sake of illustration, the energy required to recharge a typical 40 A-h, 12 V car battery, and if we further consider that the process is lossless, then the energy input is simply $40 \times 12 = 480$ W-h. If you have ever trickle charged a battery you will know that the charging process generates heat, which typically absorbs about 15% of the input power. Therefore to achieve the same level of charge we will require

$480 \times 1.15 = 552$ W-h. A battery charger connected into the main electrical supply contains transformers and rectifiers, which also generate thermal losses. It is estimated that a typical battery charger is about 60% efficient [19]. Therefore, the energy required from the 'mains' supply is $552/0.60 = 920$ W-h, that is 920 watts for an hour. But from Chap. 3 we know that almost 50% of the prime power supplied to the generation station turbines (whether fossil fuel, nuclear, hydro, solar, etc.) is lost in the electricity generation, transmission and distribution systems. Consequently prime power input at the power station in order to recharge our 12 V battery is of the order of 2 kW for an hour, or 7 MJ! This means that employing batteries for small scale storage purposes, such as to power vehicles, exerts a very expensive level of demand on primary energy sources, and as we shall see later this potentially very significant drain on renewable supplies could have a major impact on the extent to which road vehicles and in particular private cars can form part of a sustainable future even if these vehicles are electrically driven.

At the scale of storage required by the electrical power industry the unavoidable inefficiencies of battery charging and discharging are not really a problem, since the power that will be employed to recharge a battery storage plant attached to a power station would otherwise be wasted. Recharging will generally be performed when demand is low and when the wind still blows and the waves still batter the shore. In the early days of electric power generation, very large storage battery arrays were commonly installed near power stations as an essential back up for controlling demand fluctuations and for emergency systems. At that time all of the electrical power being generated and distributed was DC, which meant that the battery bank could be connected directly to the power lines. They were used to assist in ensuring economic operation of power stations and in the maintenance of supply. In so doing they were subjected to regular cycles of discharge and charge, and lead–acid batteries were harnessed for this role. Towards the beginning of the twentieth century the electrical supply industry was developing rapidly, and the advantages of very high voltage AC transmission became apparent. As a result many of the original DC stations were scrapped. However, the wholesale adoption of high voltage AC electrical power generation and transmission introduced new problems requiring the presence, at power stations, of back-up battery storage systems. Batteries were, and are, considered to be the best source of electrical supply for operating remote control switch gear, circuit-breakers, remote control equipment, and many safety and protective devices required by modern generation and distribution plants. Battery types and sizes vary considerably from station to station. For example, at the Sizewell nuclear power station in the UK the following batteries are employed [19]. Two 440 V batteries each with 224 cells are connected in parallel and are used to power emergency systems for the reactor. Each battery can supply 1300 A-h, which is equivalent to 1.144 MW-h. One 240 V battery (120 cells, 210 A-h) with an energy capacity of 50 kW-h, powers the emergency lighting, and supplies emergency power for the oil pumps. A third battery operating at 110 V (55 cells, 1200 A-h) has a capacity of 132 kW-h and is used for switching operations, while a fourth (50 V, 24 cells, 200 A-h) has a storage capacity of 10 kW-h, which is enough to power an automatic telephone exchange and station alarms. All of these battery banks are constructed from enclosed lead–acid type cells.

Clearly battery banks of moderate power have been in operation in power stations for a very long time and the technology is mature at this level of power. Renewable power stations, however, will require storage capacities that are at least an order higher than is currently the norm. Battery banks capable of storing more than 10 MW-h will be required to provide back-up storage for renewable power stations and considerable research effort is being directed towards this aim [22, 23]. Theoretically, high energy density batteries would use anodes composed of alkali metals such as sodium, lithium and potassium, which are the most reactive of metallic materials. Nickel–cadmium and nickel–zinc batteries are also being re-examined and have been shown to be potentially capable of high storage densities. Calculations, and prototype testing, suggest that alkali metal batteries are really the only source of electric power propulsion which can compete with the internal combustion engine for power delivery and range. Many combinations of reactive chemicals have been researched in the pursuit of battery solutions offering higher energy densities than the staple lead–acid version. Four chemical combinations give considerable hope that a major advance is close. These are sodium–sulphur, lithium–sulphur, lithium–chlorine and zinc–chlorine. These advanced batteries, and in particular the sodium–sulphur couple using a solid electrolyte and lithium–sulphur couple using a fused salt electrolyte, are at the prototype stage of development.

The sodium–sulphur (Na–S) battery is representative of what are termed high temperature advanced concept developments. For example, a 1 MJ capacity battery for electric vehicle applications is at an advanced stage of development at Chloride Silent Power in the UK with the collaboration of General Electric in the USA [7]. Similar battery concepts are being researched by Ford (USA), Brown Boveri (Germany) and British Rail (UK). All use a test-tube shaped ceramic container, made of beta-alumina, which is conducting to sodium ions. The tube contains molten Na in its interior (anode) and is surrounded by a sulphur melt (cathode) housed in a case, which collects the current. The operating temperature of the system is between 300 and 400°C, and the cell voltage, derived from the chemical reaction between the sodium and the sulphur to produce sodium polysulphide [24], is 2.08 V. The theoretical energy density of these batteries is about 2.7 MJ/kg, more than four times the level of the lead–acid battery. These batteries also have the additional advantage over lead–acid of better depth of discharge (~ 80%), no maintenance such as adding distilled water, and a plentiful supply of the raw materials from which they are constructed. Over 200 MW of sodium–sulphur capacity have been deployed in Japan. Generally, this has been in installations exhibiting power outputs up to 12 MW, with energy storage times of 7 hours at the rated power. In June 2006 the American Electric Power Corporation began operating the first 1 MW sodium–sulphur storage system in the USA. The acquisition of a further 6 MW of storage capacity of this type is planned.

A good example of the progress that is being made in the development of large electrochemical storage systems for electrical supply back-up, is the new battery energy storage system (BESS) at Fairbanks, in Alaska. This battery system is designed to stabilise the local grid and reduce its vulnerability to events like the blackout that occurred five years ago, on 14th August 2003, in the north eastern USA and Canada. A consortium led by the Swiss company ABB, the leading power and automation technology group, supplied and installed the BESS. At the heart of this powerful electrochemical storage system are two core components. First are the nickel–cadmium (NiCad) batteries, developed by the French company, SAFT. The 1500 ton battery bank comprises nearly 13,760 rechargeable cells in four parallel strings. Second is the convertor, designed and supplied by ABB. The convertor changes the battery's DC power into AC power ready for use in the local grid transmission system. The system is configured to operate in several distinct modes, each of them aimed at stabilising the generators if power supply problems occur. During commissioning tests in 2003 the SAFT battery and the ABB power conversion system surpassed the highest previously recorded output from a battery system by achieving a peak discharge of 26.7 MW with just two of the four battery strings operational. This makes the Alaskan BESS over 27% more powerful than the previously most powerful example, namely a 21 MW BESS commissioned by the Puerto Rico Power Authority at Sabana Llana, Puerto Rico in 1994. Although the Fairbanks plant is initially configured with four battery strings, reports [25] suggest that it can readily be expanded to six strings to provide a full 40 MW for 15 min. Recharging would take between 5 and 8 hours. The facility which occupies an area about the size of a soccer pitch, can ultimately accommodate up to eight battery strings, giving considerable flexibility to boost output or to prolong the useful life of the system beyond the planned operation span of 20 years.

Batteries are also being developed in which the electrolyte, instead of being sealed within the battery, is continually being replenished and returned to external storage tanks. These batteries are a form of fuel cell and generally exist in one of three types: zinc–bromine, vanadium redox and sodium–bromide. For example, with the zinc–bromine flow battery [26] a solution of zinc bromide is stored in two tanks. When the battery is charged or discharged the solutions (electrolytes) are pumped through a reaction vessel (battery) and back into the tanks. One tank is used to store the electrolyte for the positive electrode reactions and the other for the negative. Zinc–bromine (ZnBr) batteries display energy densities comparable with that of lead–acid types, generally of the order of 0.27 to 0.31 MJ/kg. The primary features of the zinc–bromine battery are superior depth of discharge; long life cycle; large capacity range (50 kWh), stackable to 500 kWh systems; and independence of power delivery capability from the stored energy rating.

In each cell of a ZnBr battery, two different electrolytes flow past carbon-plastic composite electrodes in two compartments separated by a microporous polyolefin membrane. During discharge, zinc (Zn) and bromine (Br) combine into zinc bromide, generating 1.8 V across each cell. This will increase the Zn ion density (each with a positive charge equal to twice the electron charge) and Br ion density (negatively charge equivalent to the electronic charge) in both electrolyte tanks [27]. During charge, metallic zinc will be deposited (plated) as a thin film on one side of the carbon-plastic composite electrode. Meanwhile, bromine evolves as a dilute solution on the other side of the membrane, reacting with other agents (organic amines) to make thick bromine oil that sinks down to the bottom of the electrolytic tank. It is allowed to mix with the rest of the electrolyte during discharge. The net efficiency of this type of battery cell is about 75%.

The development of the zinc–bromide battery is attributed to Exxon and the first examples appeared at the beginning of the 1970s. Over the years, many multi-kWh batteries of this type have been built and tested. Meidisha demonstrated a 1 MW/4 MW-h ZnBr battery in 1991 at Kyushu Electric Power company in Japan. Some multi-kWh units are now available pre-assembled, complete with plumbing and power electronics. ZBB, a company which specialises in zinc–bromine flow technology, in partnership with Sandia National Laboratories, is installing a 400 kW-h advanced BESS near Michigan in the USA.

Potential for Providing Intermittency Correction

It is clear from the extent of the literature on the subject, and on the volume of commercially sponsored propaganda on the internet, that the development of high capacity, energetic, batteries is commanding considerable interest. Improvements in energy density, efficiency and depth of discharge are the main aims of these developments and this is being pursued in a wide variety of ways. First, the well established lead–acid and nickel–cadmium batteries are being re-examined to find ways of enhancing their performance in relation to high energy storage applications. Second, more energetic materials such as sodium, lithium, zinc, sulphur and chlorine are being studied in a variety of combinations in the search for batteries offering much higher energy densities and efficiencies. Finally flow batteries, or quasi-fuel-cells, are attracting interest because they offer high capacity with power delivery being independent of the energy rating. Energy densities approaching 3 MJ/kg have already been reported, and this means that battery banks, housed in warehouses occupying no more area than a sports field, and offering storage capacities in excess of 100 MW-h, are not far off. Given the rate of technological progress in this area it may well already have happened.

4.7 Hydrogen

Storage Principle

A glance at the internet (try ‘Googling’ hydrogen) or a standard textbook of chemistry/physics will provide you with copious data on hydrogen. Suffice to say that while it makes up 75% of the known matter in the universe, hydrogen is actually quite rare here on Earth. Estimates suggest that in the surface layers of the planet, including the seas and oceans, the average concentration of elemental hydrogen is 0.14%. This makes it the tenth most abundant element coming behind titanium and ahead of phosphorus. Since it is quite reactive it exists on Earth only in combination with other elements such as oxygen in water, with carbon in methane and with nitrogen in ammonia. As a result hydrogen gas is not a readily accessible energy source as are coal, oil and natural gas. It is bound up tightly within water molecules and hydrocarbon molecules, and it takes high levels of energy to extract it and purify it. It is probably best to think of it as a carrier [28] of energy, like electricity, rather than a source of energy. On the other hand, if it were possible to fuse hydrogen molecules into helium here on Earth, mimicking the processes in the sun, then we could certainly consider hydrogen to be an energy resource. But the evidence is that fusion reactors are still very far from becoming practical sources of power in the foreseeable future, and certainly not in a time scale that is relevant to the problems of global warming.

The production of hydrogen for industrial and commercial applications has a long history [29]. Essentially there have been two main users: namely industries synthesising ammonia from hydrogen and nitrogen, as the primary ingredient in fertiliser production, and the oil industry, for high pressure 'hydro-treatment' in petroleum refineries to, for instance, convert heavy crude oils into diesel and petrol for transport usage. Global production of hydrogen is about 45 billion kilograms per year [30]. The gas is separated mainly from natural gas, oil and coal with a small percentage (4%) obtained from the electrolysis of water.

Technology Required

Natural gas, which is essentially methane (CH_4), is easily the most abundant source of hydrogen [29]. A process called steam methane reforming is used to generate the hydrogen. It is a multi-step process in which the methane is made to react with water (H_2O) at high pressure (15–25 times atmospheric pressure) and at high temperature (750–1000°C). This is done in high pressure tubes containing a catalyst (usually nickel), and the result is the formation of hydrogen and carbon monoxide (CO) [31]. In a subsequent reaction, termed the water–gas shift, the carbon monoxide is converted, in the presence of steam at between 200 and 470°C, into carbon dioxide (CO_2) and additional hydrogen [31], in one or two stages. The problem with the process is the generation of the 'greenhouse' gas carbon dioxide, a difficulty which also arises with the extraction of hydrogen from coal or oil or any hydrocarbon. Unless carbon capture, or CO_2 sequestering as it is sometimes labelled, is available, these techniques are not relevant to a sustainable low carbon future. Carbon capture is a method by which CO_2 can be 'locked up' under pressure in vast underground caverns, for example empty oil wells or natural gas wells, so that it is isolated from the troposphere in perpetuity and cannot contribute to the 'greenhouse' effect. The process is largely untried and still very much in the aegis of 'research'. Sequestration on the massive scale that would be required to generate enough hydrogen from hydrocarbons, to replace fossil fuels and provide massive storage capability, is still a very long way from being practical, but more damningly a guarantee that CO_2 trapped in this way will never seep into the atmosphere cannot be given. Studies on gas storage in geological formations [32, 33] suggest that there is 'no experimental evidence or theoretical framework' for determining likely leakage rates from such formations. The very best underground cavities have a leakage rate of 0.001% of the stored volume of gas every day [3]. For large volumes of CO_2 stored for, hopefully tens, if not hundreds, of years this is hardly negligible. Consequently, one is forced to conclude that the technique is inappropriate to a zero-emissions future. Given the impossibility of providing CO_2 storage guarantees, it also seems perverse to use this unproven carbon capture technology to justify continued burning of coal. As an engineer, and one acquainted with Murphy's laws, it is difficult not to feel that it is frankly disingenuous to claim that it is possible to take carbon, which is perfectly sequestered in deep coal seams, release it by burning it in coal power stations, and then expect to be able to put it back underground perfectly securely without at some point in the process poisoning the atmosphere.

In this section we will concentrate on electrolysis as the only sure way of generating hydrogen, to provide an energy storage medium, which is both environmentally neutral and relatively 'safe', from a massive energy storage view point. Water (H_2O) is, not surprisingly, a very common source of hydrogen. It can be 'split' by electrolysis, which is a process of decomposing water into hydrogen and oxygen by using electric current. The technology is mature and is generally used where very pure hydrogen is required. An electrolysis cell comprises five main elements. First, the containment vessel, which is not unlike a very large battery, is filled with an aqueous electrolyte (usually a dilute solution of water and potassium hydroxide). Second, an anode plate and a cathode plate are inserted into the electrolyte and are connected to an external electrical circuit, which drives current through the vessel. The electrodes are preferably made from platinum, but since this is a scarce expensive metal, the cathode is more commonly formed from nickel with a coating of platinum, while the anode is either copper or nickel coated with trace layers of the oxides of metals such as manganese, tungsten, and ruthenium to accelerate the anode interaction. Finally, the vessel is divided by a barrier layer separating the cathode electrolyte from the anode electrolyte. This layer has to be permeable to the flow of ions from anode to cathode (e.g., a proton exchange membrane or PEM), but should be impermeable to the hydrogen formed at the cathode and the oxygen formed at the anode, so that these gases can be removed separately. The chemical process can be summarised as follows: at the anode hydroxyl ions (negative) give up an electron to the electrode resulting in the formation of oxygen and water. The electron travels around the external circuit to the cathode where it combines with a potassium ion (positive). This electron flow accords with the drive current supplied by the electrolyser power source. The highly reactive potassium molecules then combine with water molecules to generate hydrogen and hydroxyl ions, which pass through the PEM to the anode, and so the process proceeds as long as current and water continue to be supplied [34].

In an ideal electrolysis cell, a voltage of 1.47 V, if applied to the electrodes at 25°C, will decompose the water into hydrogen and oxygen isothermally and the electrical efficiency will be 100%. A voltage as low as 1.23 V will still decompose the water, but now the reaction is endothermic, and energy in the form of heat will be drawn from the cell's surroundings. On the other hand, the application of a voltage higher than 1.47 V will result in water decomposition with heat being lost to the surroundings [29]. The process becomes exothermic. Clearly maximum efficiency equates to the lowest voltage that results in hydrogen and oxygen being formed. But this operating regime draws a very low current from the source and hence a very slow rate of production of hydrogen per unit area of electrode surface, which means that impractically large cells would be required to produce commercial quantities of hydrogen. As with all engineering processes a compromise is called for; in this case between efficiency and production rate. Thus, practical cells are operated at high temperature (~ 900°C) at voltages in the range 1.5–2.05 V. For example, a high temperature electrolysing cell operating at atmospheric pressure, with a power input of 60 kW, would generate 25 grams/minute or 280 litres/minute of gaseous hydrogen, together with half this amount again of oxygen (by volume) [7]. This conversion rate from input power to volume of hydrogen is calculated on the basis of negligible thermal losses. The electric current required is 40 kA for a cell voltage of 1.5 V.

Individual cells can be combined in essentially two different ways to form a hydrogen production unit. These are tank type or filter press type [29]. In electrolysers of the tank type each cell, with its anode, cathode, its own source of water, and separate electrical connections, is housed in a separate chamber; typically in the form of a rectangular container about 3 m deep by 1 m wide by 20 cm thick. These chambers are then stacked, book-like, into a unit containing about 20 cells, which are connected in parallel electrically from a low voltage, high current, busbar. The performance of an individual cell has little effect on its neighbours in this stacking arrangement, so it is a simple matter to replace faulty cells. Unfortunately, while the tank type electrolyser is electrically simple in concept, it requires the generation of very large currents. Conductors from the power supply to the tank have to be very robust, and highly conducting (usually heavy copper busbars), while massive step down transformers and rectifiers are required to supply the large DC currents. All of this drags down the efficiency of the electrolysing process. The alternative approach, termed filter press construction, is more efficient and less demanding in power supply terms. In this construction the electrodes are formed into rectangular panels, which are stacked together with suitable spacing, and with separators, like slices of bread forming a loaf. The back side of the cathode in one cell is the anode of the next cell, and the electrolysing unit will typically comprise 100 cells, electrically connected together in series. In this connection the voltages, rather than the cell currents, are additive, so that a 100 cell unit operating at 1.5 V per cell will require a supply voltage of 150 V, and a current equal to the single cell current (~ 40 kA). This is a much easier power supply requirement. However, there is a difficulty with the series connection, and that is the need for all cells to be identical, otherwise a cell can easily be overloaded and unit failure can occur because of the demise of one cell. Such a unit, producing 28,000 litres/min will be in the region of 70% efficient in converting electrical power to pressurised hydrogen gas. In size including storage tanks, it would be about 6 m high by 5 m long by 2 m wide.

Hydrogen can be stored as a liquid, as a compressed gas, and as a metallic hydride, although the third of these methods is still at an early stage of development. Liquefaction of hydrogen is a very costly process since it becomes liquid at the ultra-frigid temperature of –253°C, and storage in this form is more appropriate to transportation and transport applications (e.g., hydrogen powered buses), than to bulk storage schemes [36]. The most promising method for bulk storage of hydrogen produced from renewable energy sources is the compressed form of the gas, which can be contained in underground caverns, much in the same way as compressed air (Sect. 4.2). The very diffuse nature of hydrogen gas could result in significant leakage from such storage caverns and the technique has to rely on the fact that most rock structures tend to be sealed in their capillary pores by water [7]. Hydrogen gas at 150 atmospheres (14.71 MPa) and at 20°C has an energy content of 1.7 GJ/m^3 or 0.47 MW-h/m^3. Consequently, a suitable rock cavern with a volume of just over 1000 m^3 would be sufficient to store a very useful 500 MW-h. CAES requires 500,000 m^3 for a similar storage capacity. Clearly the high energy density in hydrogen offers very considerable storage advantages.

Storage of hydrogen in metal hydrides has also been proposed as a means of reducing storage volume. The basic concept revolves around the observation that a number of metal hydrides[7] such as $LaNi_5$, TiFe, and Mg_2Ni can absorb hydrogen at low pressures and temperatures, which can be released, with small losses, at a specific temperature and pressure. So called reversible hydrides act rather like sponges, soaking up hydrogen and storing it compactly. They are usually solids and the hydrogen can be replenished by flooding it with the gas. The process takes minutes for a tank size container with a volume of about one cubic metre. By weight the hydride sponge, when maximally soaked, contains 2% of hydrogen, although materials are being studied that can do much better than this [36].

Potential for Providing Intermittency Correction

Several large hydrogen producing and storing plants, all located near hydro-electric power stations, are in operation around the world. Currently, the highest capacity plants are in Norway, at Rjukan and Glomfjord. The Rjukan plant comprises 150 electrolysing units housed in a building the size of a large warehouse. It draws 165 MW from the nearby hydro-electric station to produce hydrogen at 27,900 m^3/hour. On the other hand the facility at Glomfjord has been installed below ground to maximise safety and to minimise visual intrusion. Storage systems of this description are considered to be superior to banks of batteries. Depending on the nature of the primary power plant, the stored hydrogen can, at periods of high demand, either be burnt in a gas turbine coupled to a generator, or be passed through a fuel cell, to produce electricity.

Hydrogen energy storage (HES) is clearly a well developed option for bulk storage and has the following advantages [30]. First, the high energy density of the hydrogen gas itself means that bulk energy storage can be achieved with relatively compact facilities. Second, such facilities are versatile in terms of storage capacity, and third, they are modular. Furthermore, charge rate, discharge rate and capacity can be treated as independent variables in the design of a hydrogen storage system. Finally, surplus hydrogen, if any, can be diverted to other applications. On the other hand, hydrogen storage is at a distinct efficiency disadvantage compared with battery and other systems. The power station-to-grid efficiency, especially if hydrogen gas turbines are employed in the chain, is less than 50%.

4.8 Capacitors

Storage Principle

For electrical and electronic engineers it is probably fair to say that capacitors are one of the most common components with which they have to deal. We have already seen in Sect. 2.4 that a capacitor in an electrical circuit in combination with an inductor forms a resonant circuit (electrical pendulum), and that such circuits are the mainstay of the ubiquitous electrical filter. In electronic circuit applications of this category, the capacitors are very small and store only tiny amounts of energy. Large high voltage capacitors tend to be used where significant amounts of electrical energy are required to be dissipated over very short time intervals, such as in testing insulators, for powering pulsed lasers, in pulsed radar, and for energising particle accelerators. Few other electrical storage systems can release, almost instantly, very high levels of power for a few microseconds or milliseconds.

The mechanism of energy storage in capacitors was touched upon in Sect. 2.4. There we addressed the notion that electrical energy, and hence electrical power, emanates from the work that has to be done in separating electrical charges of opposite sign. In addition, it has been observed, in our discussion of batteries, that if a long two wire lead is connected to the terminals of a battery the terminal voltage is transferred to the remote ends of the lead. This is because free electron charge in the conducting wire connected to the positive terminal is drawn through the battery and 'pushed' into the wire connected to the negative terminal. It is a process that occurs virtually instantaneously and is completed in fractions of a microsecond. It stops once the separation of charge at the extremity of the lead produces a voltage there that just matches the emf of the battery itself. Now, if the wire from the positive battery terminal is attached to a large flat metal plate or electrode, while the wire from the negative terminal is connected to a second plate of equal size, which is close to, and parallel to, the first plate, forming a metal–air–metal ‘sandwich’, then current will flow through the battery for much longer. The reason for this is actually quite simple. As before, the criterion for the process to stop is that the voltage at the plates must equal the battery emf. But for this large parallel plate structure, where do we judge that the voltage occurs? Is it at the edges of the plates, in the middle of the air gap or at some other point in the air gap? Well it has to be the same everywhere otherwise the process cannot be said to have stopped. If there is a voltage gradient, between any two points in the parallel plate structure, charge will continue to flow in the conducting plates until no voltage gradients exist. The amount of charge that has to be transferred from the positive plate to the negative plate, through the battery, to achieve this steady state is the product of the voltage and the charge storage capacity of the parallel plate system, termed the capacitance [37, 38].

Technology Required

For a parallel plate capacitor the capacitance in farads is easy to compute, being proportional to the area between the plates and inversely proportional to the separation distance [38]. For example, 1 m × 1 m square plates in air, separated by a distance of 1 cm, exhibit a capacitance of 0.88×10^{-9} farads, or 0.88 nF (n denotes nano). The energy stored in the capacitor can be determined by computing the work that has to be done to separate the plates by 1 cm, against the force of attraction between the positive charge on one plate, and the negative charge on the other. (Actually keeping the plates separated requires a mechanical structure to prevent them moving together.) This leads to the result that the stored energy in joules is given by half the capacitance multiplied by the voltage squared [37, 39]. Therefore, if our square-plated air spaced capacitor is charged from a battery bank generating 10 kV (say) the energy stored in the electrostatic field formed in the 1 cm gap, will be 0.044 joules with a density of storage of 4.4 J/m^3. In bulk energy storage terms this is a paltry amount and it would take many barnloads of capacitors to get to the MW-h level!

Capacitor energy storage potential can, however, be enhanced very significantly by intelligent use of dielectrics in the electrode gap, instead of air. This, by the way, is a much simpler way of keeping the electrodes separated than a mechanical restraining structure. Increased storage capability occurs because capacitance is proportional to the relative permittivity, or refractive index, of the material separating the electrodes [38]. Actually, it is slightly more complex than this because the capacitance is also significantly influenced by whether or not the material is non-polar or polar (the choice still exists – unlike the Arctic which will soon be non-polar everywhere!), and whether or not it is easily ionised. In insulating materials, or dielectrics, all orbiting electrons are tightly bound in covalent bonds (electron sharing) to the fixed positive nuclei, and the material (e.g., glass or mica) is usually dense, hard, and brittle. For most solid dielectrics, atoms comprise a cloud of electrons orbiting a fixed nucleus, and the centre of the cloud is coincident with that of the nucleus. Think perhaps of a hollow globe (electron cloud) with a tiny lead weight (nucleus) at its centre, held there by radial spokes. The material is said to be non-polar when the charges in all atoms are symmetrically distributed in this way. Now, when such a material is placed between the plates of a capacitor that has been charged to a voltage greater than zero, each atom will be immersed in an electric field. This field will tend to pull orbiting electrons towards the positive plate. Returning to our globe analogy, if the spokes were not quite rigid, by replacing them with stiff rubber bands, the centre of the globe and the centre of the lead weight can no longer be co-located except in a zero gravity chamber. In the absence of such a chamber the lead weight will be pulled by gravity towards the base of the globe, so that it is no longer centred in the globe. Note that this off-set would persist even if the globe was spinning at constant speed. The off-set is wholly due to the gravitational field. In electrical terms, if the lead weight represents the positive nucleus of an atom and the globe represents the orbiting electron cloud, the displacement of the centre of positive charge from the centre of negative charge occurs as a result of the electric field between the capacitor plates (instead of gravity). Each atom is described as a dipole, and the material as a whole is now said to be polarised, if all of the dipoles are aligned in the same direction. The resultant charge separation in the material, which is in the opposite direction to the charge on the electrodes, has the effect of reducing the field between the plates, and more charge has to be supplied to the electrodes, from the battery, or power source, to maintain the voltage. Given that, at constant voltage, capacitance is proportional to charge as we have already observed, it is evident that the insertion of the dielectric has a similarly direct influence. In fact capacitance, for a device containing a simple non-polar dielectric, increases in direct proportion to its relative permittivity, as we noted earlier. For example, if the air gap in our parallel plate structure were filled with glass with a relative permittivity of about 10, its capacitance would increase to 8.8 nF. This is still too small to be interesting in bulk storage terms, and in any case, glass filled capacitors, unless they are very small, are highly impractical because of the rigidity, fragility and density of glass. For the same structure size, even higher capacitance is possible by employing an exotic ceramic such as barium strontium titanate, which has a relative permittivity of $\sim 10{,}000$. Unfortunately such materials are extremely expensive because of there scarcity, and are consequently somewhat irrelevant to the search for a solution to the bulk storage of electricity using capacitors.

Polar materials are slightly more promising in offering high permittivity from non-exotic compounds. In such substances molecular dipoles are already present in the isolated, neutral, form. The most abundant of these is water, in which the H_2O molecule is asymmetric. While each hydrogen atom is strongly bonded by sharing electrons covalently with the oxygen atom, the electron cloud of the molecule tends to favour the oxygen nucleus leaving the hydrogen nuclei exposed. As Angier [40], in *The Cannon* puts it: the molecule 'is best exemplified by the stridently unserious image of Mickey Mouse ... with the head representing oxygen, the ears the two hydrogen atoms covalently linked to it'. Because of the asymmetry 'the ears of the Mickey molecule have a slight positive charge … the bottom half of the mouse face has a five o'clock shadow of modest negative charge'. In a mass of water the 'chins of one molecule are drawn to the ears of another' so that water molecules cling together just enough to give it its liquid properties. This dipole bond, or hydrogen bond as it is more commonly called, is only about one tenth as strong as the covalent bond binding the 'ears to the head'. Electrically, the dipolar water molecules are very susceptible to electric field, so that when capacitor plates are immersed in pure water the 'ears' are attracted to the negative plate, and the 'chins' to the positive plate. The dipoles become aligned and the water becomes polarised. This happens generally at a much lower voltage than for a non-polar material. Thus pure water has a high relative permittivity, tabulated as 81 for distilled water. However, this is still not enough to produce energy density levels that are significant in bulk storage terms.

The remaining possibility is electrochemical capacitors. In this category electrolytics are the most well established embodiment. High capacitance is achieved in electrolytic capacitors by introducing an electrolyte into the space between the metal electrodes. In this type of capacitor ions in the electrolyte provide a mechanism for conduction current flow and the electrolyte can thus act as one of its plates. High capacitance is procured, not by employing a polarising effect in the electrolyte, but by separating it from the second plate by an extremely thin oxidised insulating layer on this electrode. Aluminium electrolytic capacitors are constructed from two conducting aluminium foils, one of which is coated with an insulating oxide layer, separated by a paper insert soaked in electrolyte. The electrolyte is usually boric acid or sodium borate in aqueous solution together with various sugars of ethylene glycol, which are added to retard evaporation. The foil insulated by the oxide layer is the anode, while the liquid electrolyte and the second foil act as cathode. This stack is then rolled up, fitted with pin connectors and placed in a cylindrical aluminium casing. The layer of insulating aluminium oxide on the surface of the anode acts as the dielectric, and it is the thinness of this layer that allows for a relatively high capacitance in a small volume. The aluminium oxide layer can withstand an electric field strength of the order of 10^9 volts per meter, so relatively high voltages can be applied to the device without incurring catastrophic breakdown. This combination of high capacitance and high voltage gives the electrolytic capacitor its high energy density. For example, if we insert our 1 m square plate capacitor into an electrolyte so that the electrolyte is separated from the positive plate by a 10 μm thick insulating layer, the capacitance becomes 5 microfarads (5 μF) assuming that the insulating layer has a relative permittivity of 6, which is typical of a metal oxide. The energy stored at 10 kV is now 250 J, or about 25 kJ/m^3. This is beginning to approach levels that are significant in bulk storage terms. Research into electrochemical capacitors (EC), which store electrical energy in two insulating layers when oxide coated electrodes are separated by an electrolyte (electric double layer, EDL), indicates that the separation distance over which the charge separation occurs can be reduced to a few angstroms (1 angstrom = 0.1 nm). The capacitance and energy density of these devices is thousands of times larger than electrolytic capacitors [41, 42]. The electrodes are often made with porous carbon material. The electrolyte is either aqueous or organic. The aqueous capacitors have a lower energy density due to a lower cell voltage but are less expensive and work over a wider temperature range. Furthermore, electrochemical capacitors [43] exhibiting higher voltage and higher energy density limits than is currently available appear possible if polymer-based insulating layers can be formed with dielectric constants that can be increased without compromising thermal and mechanical properties or the ability to clear defect sites. Sophisticated computer modelling at the molecular level is employed to devise suitable compounds.

Potential for Providing Intermittency Correction

Compared with lead–acid batteries, EC capacitors tend to have lower energy densities but they can be cycled tens of thousands of times and are much more powerful than batteries because of the speed at which they can be discharged (fast charge and discharge capability). The current state of the art is that while small electrochemical capacitors for energy storage application are well developed, larger units with energy densities over 20 kW-h/m^3 (72 MJ/m^3) are still under development. Capacitor banks in warehouses each occupying a modest area of about 1000 m^2 could be capable of storing 20 MW-h or more, in the not too distant future, if a serious, well funded, commitment were to be made to advance the technology to production level.

4.9 Superconducting Magnets

Storage Principle

In Chap. 2, Sect. 2.4, you may recall the observation that when charge is in motion (thus producing a current) it possesses additional energy, not unlike the kinetic energy of a moving mass in a gravitational field, and that this energy is stored in a magnetic field. For current-carrying conductors, the relationship between magnetic field and current flow can be determined using one of the most fundamental electrical laws, namely that due to Ampere. For a long straight conductor it yields the result that the magnetic field intensity, which describes circular paths centred on the wire, is proportional to the current and inversely proportional to the distance from the wire [44]. On the other hand, for a current-carrying coil, which has a large length to diameter ratio, the magnetic field intensity threading through the centre of the coil is proportional to both the current and the number of turns, and inversely proportional to its length [44].

To establish the magnetic energy in a coil clearly we have do work, in accordance with the first law of thermodynamics. We have to do work, because in raising the current from its initial value (probably zero) to its final value, a changing magnetic field is being experienced. But as we have seen in Sect. 2.4, changing the magnetic field produces a force (Faraday affect) that is trying to resist the current increase. The force is generally termed the back emf. This back emf is independent of whether or not the coil is superconducting. Having determined the back emf it is then possible to integrate the work done per unit charge over time and thence compute the energy stored in a coil of known dimensions. The results are essentially the dual of the energy equations for capacitance. If the inductance of the coil is known, which it usually is, then the energy stored in it is equal to half the inductance multiplied by the current squared [45]. Let us consider applying this to a coil of dimensions suitable for substantial energy storage. In electrical engineering terms it will be very large, at a guess something like 5 m long and 0.3 m in diameter. Its inductance, if air filled ($\mu_0 = 4\pi \times 10^{-7} H/m$), will be 5.7 H, and for a current of 500 A (say) the energy stored in it will therefore be 0.71 MJ. This gives an energy stored per unit volume of ~2 MJ/m^3. This is a considerably more promising level than for basic capacitor systems (1–2 kJ/m^3), but is not particularly impressive when compared with the storage density in a battery.

What is required to improve energy storage, is the ability to drive much more current through the coil. Normally this is not possible because of coil resistance and excess heating due to joule loss in the metal (copper, aluminium) forming the coil. However, with supercooled coils this limitation is greatly relaxed. When supercooled, some conductors are able to carry very high current and hence high magnetic fields with zero resistance, if the temperature is low enough. Such metals are termed superconductors. Superconductivity occurs in a wide variety of materials, including simple elements like tin and aluminium, various metallic alloys and some heavily-doped semiconductors [46]. Superconductivity does not, however, occur in copper, nor in noble metals like gold and silver, nor in most ferromagnetic metals. As an example of the superconducting temperature threshold, aluminium is superconducting below 1.175 K, which in Centigrade terms is –271.825°C.

That engineers are, today, pursuing the notion of storing large amounts of electrical energy in massive supercooled superconducting coils is hardly surprising. With zero resistance, losses will be negligible, and such a system offers the possibility of very efficient storage. Since it stores electrical energy directly, it can, not unlike capacitor storage, be linked straight into the electrical supply system through suitable switching arrangements and DC/AC convertors. When a superconducting coil is attached to a DC supply the current in the coil grows, much as for a conventional coil, until it becomes limited by the supply. The primary difference, from an un-cooled device, is that all of the power supplied by the DC source is converted into stored energy. None is wasted in heating the coil. Once the maximum DC current is reached, the voltage across the terminals of the zero resistance superconducting coil, drops to zero. The current keeps flowing with no input from the supply. This is not unlike a flywheel in a vacuum, and on frictionless bearings, which will continue to spin in perpetuity unless braking is applied. For the fully 'charged' coil the magnetic energy can be stored as long as required with no loss to the generating system. However, problems do have to be overcome with superconducting magnetic energy storage (SMES) systems. These can be summarised as follows:

- effective and reliable very low temperature refrigeration;
- effective shielding to contain stray magnetic fields;
- accommodating the high mechanical forces generated during charging and discharging; and
- protection against unexpected loss of superconducting properties.

Technology Required

Practical superconducting coils are currently formed from multi-cored wires containing filaments made from niobium/titanium (NbTi) or niobium/tin (Nb_3Sn) compounds [7]. In cross-section the wire is divided by aluminium radials into eight sectors, and this structure is contained within a thin cylindrical sheath also made of aluminium. The eight sectors are filled with a super pure aluminium matrix for stabilisation and the superconducting filaments are located in a circumferential ring just inside the sheath [46]. The superconducting filaments are mainly formed from niobium/titanium compounds, which are relatively easy to manufacture. Such a compound with 47% niobium and 53% titanium has a critical temperature of 9.2 K, below which it is superconducting. At zero degrees it can theoretically conduct a current of 10,000 A/mm^2. The design of the cable with its stabilisation matrix of pure aluminium [47], ensures that if the superconducting filaments become normally conducting for whatever reason, current will flow with lower density in the aluminium, thus avoiding cable, and hence coil, destruction through overheating.

Storage coils for SMES systems generally fall into one of three categories. These are single circular cylindrical solenoid, series connected flat coils mounted coaxially, and series connected single coils wound on a torus. Solenoids are used widely in electrical engineering and electronics to provide magnetic storage for inductors and transformers, and it is well known that to minimise leakage and interference from stray magnetic fields the length to diameter ration (κ) of the solenoid should be large ($\kappa >> 1$). However, in SMES terms long solenoids make poor use of the superconducting material, which has to be used sparingly. Consequently, flat solenoids with $\kappa < 1$ are preferred. Because leakage magnetic fields are high, series connected, and coaxially aligned, flat solenoids, are inevitably subjected to very high radial and axial forces generated by the Lorentz effect (see Sect. 2.4). Mechanical stiffening and magnetic shielding is necessary to compensate for this. Coils wound on a torus behave much like a long solenoid displaying low stray magnetic field levels, but they are expensive in their use of superconducting wire. Shielding requirements are low but strong radial Lorentz forces require mechanical reinforcing.

SMES systems for use in power station support roles suggest the need for coils carrying currents in excess of 500 kA. At these kinds of currents the Lorentz forces within the coil are enormous, enough to burst or crush the coil, depending on its design. The design of such coils is therefore dominated by the need to counteract these forces. Self-supporting structures to hold the coil together against the disruptive forces would make SMES much too expensive to implement. The recommended and generally accepted solution entails placing the windings in underground circular tunnels cut into suitable bedrock. The tunnel is required to house the coil obviously, but also, the anchors to the bedrock, the liquid helium jacket, the vacuum jacket and the refrigeration system. A typical tunnel would be about 100 m in diameter and perhaps 10 m high and 10 m wide, which is small by mining standards. A coil with 2675 turns, cooled to 1.80 K and carrying a current of 757 A is estimated to be capable of storing 3.6×10^{13} J (36 TJ) of magnetic energy [7]. Studies involving computer simulations can give some idea of the potential for SMES. For example a Wisconsin University study [48] shows that a three coil system, in three 300 m diameter, circular tunnels, arranged coaxially at three different depths of about 300 m, 350 m and 400 m, could store 10,000–13,000 MW-h of magnetic energy. Maximum power outputs range from 1000 to 2500 MW with discharge times of 5 hours to 12 hours. Coil currents range from 50 to 300 kA. Efficiency is predicted to be of the order of 85–90% with primary losses being in refrigeration (20–30 MW), and in conversion from DC to AC, resulting in an added loss of about 2% of the delivered power.

The start of research and development work on SMES is generally placed in the 1970s and is attributed to companies in quite diverse locations such as France, Germany Japan, Russia, UK and USA, with the most significant developments taking place in Japan, Russia and the USA. The High Temperature Institute (IVTAN) in Moscow has been engaged on a number of SMES projects since 1970, and since 1989 this research has been sponsored by the Russian State Scientific 'High Temperature Superconductivity' Programme [7]. By the mid-1990s IVTAN had installed, in its experimental campus, an SMES system with a storage capacity of 100 MJ and an output power of 30 MW [49]. It provided back up power to the nearby 11/35 kV substation of the Moscow Power Company. An SMES system has also been designed by the Los Alamos National Laboratory and a commercial version has been built for the Bonnyville Power Company in the USA [48]. This device, with a 1.29 m diameter and 0.86 m high superconducting coil, was rated at 30 MJ and was capable of delivering 10 MW at ~ 5 kA.

Potential for Providing Intermittency Correction

The path from prototype development to full scale implementation of a technology is often a precarious one, and SMES represents a technology that requires the solution of very complex scientific and engineering problems. Success in 'rolling out' this technology in the foreseeable future will take a very major commitment of manpower and funding, but decisions need to be made now rather than tomorrow! Should this happen, large underground systems capable of storing up to 12 GW-h using currents of 50 kA, in 300 m diameter supercooled coils, are predicted to be a realistic outcome. Such a facility could produce 2.5 GW of power for ~5 hours, which would undoubtedly represent substantial load-levelling and stabilisation potential for renewable power stations.

4.10 Nuclear Back-up

The expending of serious and meaningful research and development time and money is clearly needed to bring MES systems up to a level which matches the progress that has been made in establishing the technology for renewable power generation. From a strictly engineering standpoint, it is difficult not to accept that the lag between storage and generation can be ameliorated by employing nuclear power generated electricity, to furnish base load when MES systems are found wanting. However, as is very well known, nuclear power generation is controversial for a host of reasons, many of which are spurious [50], particularly those relating to the environment. Safety issues are a cause for concern as we shall see.

A nuclear power station generates electricity in a manner that is very little different from a conventional coal powered station, except in the way in which the steam is produced to drive the steam turbines (Sect. 2.5). In a conventional nuclear station it is a byproduct of the process of cooling the reactor – a device in which nuclear chain reactions are initiated, controlled, and sustained at a steady rate. In a nuclear bomb, by contrast, the chain reaction occurs in a fraction of a second and is uncontrolled thereby causing an explosion.

The internet is replete with tutorials on nuclear fission, and there are large numbers of educational text books that make a good attempt at rendering the topic digestible [51]. The subject will merely be 'skated over' here, to clarify what the supplying of electricity from nuclear fission entails. In atoms that have large nuclei, such as uranium-235, plutonium-239 or plutonium-241, the forces binding the protons and neutrons together are stretched to their limits, and the material is said to be fissile: i.e., it is not too difficult to cause it to split. This can happen if such a nucleus absorbs a neutron: it succumbs to nuclear fission. The original heavy nucleus divides into two or more lighter nuclei thereby releasing kinetic energy, gamma radiation and free neutrons; collectively known as fission products. In a suitable containment vessel for the radioactive source material, and under the right circumstances, a portion of these neutrons may subsequently be absorbed by other fissile atoms and trigger further fission events, which release more neutrons, thus initiating a chain reaction. The tricky part from an engineering viewpoint is controlling the chain reaction. Fortunately it can be kept in check by using so called neutron moderators, which have the effect of changing the proportion of neutrons that will go on to cause more fission. Commonly used moderators include regular (light) water (75% of the world's reactors), solid graphite (20% of reactors) and heavy water (5% of reactors). Clearly, moderating the rate of fission has the effect of increasing or decreasing the energy output of the reactor.

The reactor core generates heat in a number of ways. First, the kinetic energy of fission products is converted to thermal energy when these nuclei collide with nearby atoms of the coolant – often water but sometimes a liquid metal. Second, some of the gamma rays produced during fission are absorbed by the reactor in the form of heat; and third, heat is produced by the radioactive decay of fission products and materials that have been activated by neutron absorption. This heat associated with radioactive decay will remain for some time even after the reactor is shutdown. The heat is carried away from the reactor by the coolant and is then used to generate steam. Most reactor systems employ a cooling system that is physically separated from the water, which will be boiled to produce pressurised steam for the turbines. An example of such a reactor is the pressurised water reactor. But in some reactors the water for the steam turbines is boiled directly by the reactor core, for example in the boiling water reactor. Needless to say, the heat power generated by uranium in a nuclear reaction is of the order of a million times greater than that of the equal mass of coal. Power outputs from installed reactors around the world range from a few megawatts (MWe) to just over a gigawatt (GWe) of electrical power. Energy capacities are typically at the TW-h level. Available statistics suggest that in 2008 the installed capacity of nuclear generators around the world was close to 0.8 TW. It is possible that a further 1 TW of electrical power could be generated by nuclear fission by 2030, but it would take an unprecedented build rate to do so – probably about two 500 MW stations per week. Unfortunately at this rate of build and operation, readily accessible reserves of uranium run out at about 2040 [52].

The future for nuclear power is immensely difficult to gauge. Some countries such as the USA are resisting moves to renew their aging power stations while others, such as France, are enthusiastically growing their nuclear real estate. Around the world the issues of new build and renewal in the nuclear power industry are political 'hot potatoes'. The link to bombs, the dangers of radioactivity, waste disposal and safety, all feature in the debate, which is 'spun' beyond rationality by the degree of polarisation demonstrated by the protagonists. For example, the Chernobyl disaster in the Ukraine, happened sufficiently long ago (1986) for dependable statistics on the effects of a nuclear accident to accumulate. However, in the view of the pro-nuclear lobbyists, the 75 deaths of mainly firemen and rescue workers pale into insignificance by comparison with deaths in coal mines, and deaths on the roads. They are of little consequence it seems. On the other hand the Greens tend to emphasise that, in the Ukraine, which suffered most from the fall-out, unusually high rates of thyroid cancer occur in children, and radiation sickness in the general population is still atypically high long after the meltdown. Needless to say, the pro-nuclear groups dispute the evidence for this.

Given the extreme level of the hype and spin that surrounds this issue, the only way to get a reasonably rational evaluation of the role of nuclear fission as part of any future sustainable energy supply system, is simply to take a very pragmatic engineering approach. The first point that needs to be made, which cannot be disputed, is that nuclear fission involves controlling a continuous explosion. As such, a nuclear reactor is only conditionally stable, requiring a very complex and elaborate computer controlled sensing and monitoring system to keep the chain reaction within safe limits. The operation is not unlike flying an intrinsically unstable high speed military aircraft of the stealth variety perhaps. If anything goes wrong, the aircraft can become impossible to fly. As experienced engineers well know, with complex systems, if anything can go wrong it usually will eventually. On the other hand, all of the renewable power systems described in Chap. 3, are essentially stable, like a glider. Something may go wrong, but a 'soft landing' is unlikely to be impossible.

Second, it cannot surely be disputed that the mineral uranium is a finite resource. It exists in various forms in the Earth's crust and oceans. It is estimated that 5.5 million tonnes of uranium ore reserves around the world are in the economically viable category [53]. Much more uranium, classed as mineral resources with some prospect for eventual economic extraction, is known to exist and is estimated to amount to 35 million tonnes. Furthermore, an additional 4.6 billion tonnes of uranium are estimated to reside in sea water, but these non-ore reserves are largely irrelevant to nuclear power generation between now and 2030. It is estimated [53] that by 2025, world nuclear energy capacity is expected to grow to about 500 GWe. This will raise annual uranium requirements to between 80,000 and 100,000 tonnes. This, in turn, means that we can sensibly rely on uranium as a viable source of energy for about 55 years. A major expansion in the build rate of nuclear fission power stations (we would need 50,000 stations of typically 500 MW capacity to meet a projected 'business-as-usual' demand of 25 TW by 2050) would seriously shrink the time duration to a world depleted of exploitable uranium.

Finally, radioactivity, which is a particularly troublesome by-product of the nuclear industry, is undeniably harmful to biological cell structures, and hence to living creatures. The debate between the pro- and anti-nuclear camps is generally fixated on the level at which this harm arises. For humans a radiation dose is fatal if it exceeds 3 sieverts (Sv). For gamma radiation this implies tissue absorbing an energy level of 3 J/kg, or 3 W for one second in a kilogram of tissue. To put this in perspective, humans naturally absorb 0.01 Sv from background radiation over a period of about 3 years. So to what extent would a nuclear accident of Chernobyl proportions irradiate a local population? The published figures indicate that during the 50 years since the accident the radiation exposure from this event amounted to 930,000 person-Sv – i.e., on average 930,000 people could have experience a 1 Sv dose; some perhaps more some perhaps less. Not enough to cause a large number of deaths but certainly enough to cause a great deal of ill health for a lifetime over an extensive area of countryside. Clearly a nuclear event is quite unlike a disaster such a gas explosion in a mine, or the bursting of a hydroelectric dam, which are relative local and short term in their effects – a nuclear incident is potentially of global significance because of fallout.

Statistically it is quite clear that as nuclear power stations grow in number and become more widespread, the more likely is it that a major incident will occur, and this increasing risk must be acknowledged in the planning process. Even staunch believers in the benefits of nuclear technology would find it hard not to be disturbed by the idea of coast to coast nuclear power stations to counteract global warming. Certainly, engineers with some knowledge of the laws of Murphy would have great difficulty in viewing this prospect with any equanimity.

As the current inhabitants and stewards of the Earth we should surely be extremely cautious about contemplating the nuclear route towards securing an alternative primary energy source to replace fossil fuels? Certainly from an engineering perspective a much wiser course for mankind would be to plan for the modernisation of the industry, while maintaining nuclear provision at roughly the level it is now in 2008, in order to minimise the potential for a severe accident and to maximise the useful life of exploitable uranium. The aim should be to employ this contained nuclear industry as a base load supplier of electrical power, during the transition to a future resourced by renewable power.

4.11 The Ecogrid

This gentle trawl through the highways and byways of massive energy storage has, if nothing else, clearly underlined one stark fact. Namely this – because of the piecemeal approach to renewable energy, there has been a negligent lack of commitment to MES from governments around the world, and a serious absence of investment from business. Consequently, a significant range of technologically challenging solutions to the renewable energy storage dilemma still exists, despite the best efforts of engineers and scientists. This is hugely undermining meaningful development of this key technology, and it is imperative that the unfortunate emphasis on power generation at the expense of energy storage should be rectified, if a successful transition to a sustainable future based on renewable resources is to be achieved. Prudence therefore suggests that some reliance on a nuclear power 'back stop' will be unavoidable, in the transition to sustainability.

Several times in this and the previous chapter, we have alluded to an unavoidable feature of renewable power generation, which distinguishes it from fossil fuelled power, namely that the supply is intermittent, particularly from wind, wave and solar sources. Engineers in the present day electrical supply industry, which is mainly reliant on fossil fuels, tend not to be too exercised by the possibility of the power generation being unreliable. Consequently they are inclined to compare fossil fuel power stations, which are normally linked directly to the grid, with renewable power stations presumed to be organised to do the same. In addition, they are inclined to view storage facilities as 'load'. Not surprisingly, simplistic comparisons of this ilk tend to cast the renewables option in a very unfavourable light. In fact, it is actually quite misleading to view renewable power generation as a direct replacement for our fossil fuel based system. Based on purely engineering considerations, an appropriate strategy, for the replacement of fossil fuels with renewables for the generation of electricity, would be unlikely to adopt technical solutions, which have underpinned conventional power generation. Rather, by developing and installing MES systems at the same pace, or more quickly, than the renewable power stations, it becomes no longer meaningful to consider the grid as being supplied primarily from the power generators. It becomes more appropriate to view the transmission network as being fed primarily from MES systems, which are being continually 'topped up', as and when, the stations are delivering power. Some base load supply from 'back-up' renewable power stations is also assumed to be available. Nuclear fission is regarded here as a MES source. With this scenario, provided there are enough renewable power stations of different types to ensure stored energy is always available, intermittency, in principle, no longer presents a problem. The grid is 'buffered' from intermittent power generators by the MES systems, the range of sources supplying the grid and the wide geographical spread of these sources. According to Ter-Gazarian [7]:

> The problem of integrating energy storage into a power system is one of the most interesting ones facing power utilities today. In any scenario of power system expansion there needs to be efficient storage of generated electricity. It is equally essential for nuclear or coal powered plants and for large scale exploitation of intermittent renewable resources.

So how will this MES buffered, renewables based, electricity supply system operate? Table 4.1 can help in making some informed guesses. But before we consider the implications of the comparisons presented in the table, it will be helpful to recall the simple electrical circuit, which we introduced in Sect. 2.6 to illuminate the operation of the grid. We can use the same simple circuit concept to illustrate the operating principles of a massive electricity storage, generation and transmission system, which henceforth will be referred to as the ecogrid. The pair of connecting wires of our elementary circuit (the grid) are connected to rechargeable batteries (MES), electric bulbs (loads/consumers), while the batteries are also connected to clockwork charging devices (renewable electric power stations). As before, one interconnecting wire will be routed to the positive terminals of the array of batteries, while the other will be connected to the negative terminals (usually earthed). Interspersed around the circuit (which could be a single loop or many interconnected loops), light bulbs (loads) are connected in parallel with the loops (i.e., between the positive and negative wire) through a simple switch. For the moment we can forget the switching arrangement introduced in Sect. 2.6 to represent AC operation. For this 'renewables' grid, or circuit, there is an additional input, namely that provided by the clockwork powered battery chargers. In this case, the positive terminals of the battery chargers are attached to the positive circuit loop while the negative terminals are earthed. Clearly the load on the circuit (grid) varies with the number of bulbs that are switched on. If all bulbs are 'on' we have peak consumption, which should normally, in grid simulation terms, be of short duration. During this period enough power must be available from the batteries plus the chargers to satisfy demand (keep the bulbs lit). If all the chargers are fully wound up, and all the batteries are fully charged, this will not be a problem, assuming that battery capacity and accumulated charger power have been properly established on the basis of the demand statistics – a calculation with which power engineers are very familiar. However, to simulate renewable power generation, which is intermittent, we really have to assume that not all chargers will be wound up and delivering charge all of the time: that some clockwork mechanisms will have wound down, or are about to wind down, or are about to be wound up. Again, statistical assessments of 'down-time' in addition to demand will be required, to ensure enough chargers, and enough storage capacity, are always available to meet demand. Finally, when the demand is negligible (all bulbs 'off') the batteries become loads, and in this case, there should always be enough storage capacity to absorb the power from those chargers, which are currently fully wound, thus avoiding power wastage. Once more, statistical calculations will be required to get an optimum balance between demand, battery capacity, and available charging power. If the circuit is large and there are enough chargers and rechargeable batteries, and if the load variation and the clockwork mechanisms are not wholly random, this balance will not be too difficult to compute. In reality, of course, the situation is much more complicated than this [7], but the general idea that the circuit model gives, of a potentially secure and dependable renewables based grid, is sound enough.

A glance at Table 4.1 quickly informs us that MES systems differ hugely in their capabilities and characteristics. The range is useful because systems can be employed in different roles [7] in the ecogrid. For example, flywheel (FES), capacitive (CES) and magnetic (SMES) storage systems exhibit rapid response times and can be recharged quickly. They are also site independent and efficient. Consequently, they lend themselves to primary storage roles close to the consumer. They could be housed on industrial estates in warehouse like structures, or on electricity distribution substations, where environmental issues are hardly likely to arise. Safety concerns such as flywheel explosion, high voltages and stray magnetic fields, should not be difficult to contain. On the other hand, large battery complexes (ECES) and hydrogen storage (HES) systems are much less likely to be tolerated close to large populations, because of safety issues. These systems are more suited to location close to renewable power stations, particularly wind farms and solar farms, as second stage storage. Since charging and discharging will be less frequent in these roles, low efficiency becomes less of a concern.

Finally, the large inflexible systems, namely pumped hydro (PHES), compressed air (CAES) and thermal (THES) will generally be independently sited at locations where suitable geological formations exist. They will tend to be delegated the role of providing third stage back-up storage, to cover unusually high demand or unusually poor levels of power generation. This means that their slow response times, long recharge times and moderate efficiency, are unlikely to be problematic. Today, PHES already undertakes an operational role on national grid systems, which is not too dissimilar to this.

To operate dependably, basically by incorporating sufficiently large numbers of both power stations and storage facilities of varying types, the ecogrid will have to cover an extensive area – at least continental in size. It is possible that for additional security of supply, continental ecogrids may also be linked together. As a result, very long transmission distances of more than 500 miles – much longer than is common on national grids – are inevitable. Transmission over these distances can only be done efficiently, as we saw in Sect. 2.6, by employing high voltage direct current (HVDC) techniques, perhaps using superconducting intercontinental links. On AC lines expensive reactive compensation techniques are increasingly required over distances greater than 100 miles, and AC grids will remain confined to national and local supply systems. In is pertinent to note here that a European quasi-ecogrid, with minimal consideration for the storage element, but based on an HVDC grid system interconnecting geographically remote and widely dispersed renewable power stations, of various types, is already being discussed [54].

The disadvantage of HVDC is that voltage is not transformable from one level to another: from say 400 kV for HVDC transmission to 33 kV for the local grid. Consequently there is a need for rectifiers and inverters to convert AC to DC and vice versa. The 'inverting' substations to link the HVDC lines to the AC lower voltage grid system will be more complex and costly than conventional AC substations, and will introduce additional losses into the system (0.6% as against 0.2% for a conventional substation). However, the power savings accruing from adopting HVDC transmission over long distances will more than compensate for this loss.

While technologically the ecogrid is viable, it can be realised only with sufficient determination by the international community to create it and with enough resources made available to implement it. Unfortunately, there is a distinct possibility that it will be dismissed as being unrealistic, both economically and politically. It could not, for example, 'get off the ground' in a politically divisive world. This complex system, which is intended to span continents, is potentially highly vulnerable to unsympathetic humans, and without cooperative nations providing maintenance and security guarantees, and agreeing power rationing mechanisms, it could not function. But that, depressingly, is a matter for politicians.

Notwithstanding these political hurdles, it is instructive to continue pursuing the engineering logic. As we have shown, in this and the previous chapter, all of the enabling components for creating a secure, reliable, electricity supply system from renewable energy resources already exist, in various stages of development, from conceptual to commercially available (see Table 4.1). But, and it is a big but, to realise an ecogrid delivering 14 TW from renewables alone by 2050 (see Table 3.1) we will require to construct renewable power generating complexes at the rate of 870 MW per day, given that almost 1 TW of renewable power is already on-stream. This seemingly impossible task involves the development, construction and commissioning of power collectors, generating plant, inverters, transformers, low loss HVDC lines, and storage facilities. If we presume that an average renewable power station including storage will typically be 250 MW in size, we will need to build by 2050 at the latest, 52,000 stations strategically located around the world at the rate of three a day, in addition to the HVDC grid itself. Leaving all other considerations aside, treating it purely as an engineering project, it is just about possible to suggest a plan of action to achieve this rate of build, and we shall consider it in the next chapter. The possible political and commercial implications of what will be discussed therein, are alluded to, but without too much comment, in order to maintain a clear focus on engineering solutions and how they can be prosecuted.

Table 4.1 Massive energy storage – Comparative performance and availability

Energy storage system	PHES[1]	CAES[2]	FES[3]	THES[4]	ECES[5]	HES[6]	CES[7]	SMES[8]	NBL[9]
Power rating (MW)[10]	100–4000	100–300	10–100	1–10	10–50	50–500	10–100	10–30	100–1000
Reported capacity (MW–h)[11]	90,000	6,000	1–2	50	1000	500	1–2	1–2	N/A
Response time[12]	<min	sec	<cycle	>min	<cycle	sec	<cycle	<cycle	slow
Recharge time[13]	Slow (>10hrs)	Medium (5–10hrs)	Fast (<5hrs)	Slow (>10hrs)	Medium (5–10hrs)	Medium (5–10hrs)	Fast (<5hrs)	Medium (5–10hrs)	N/A
State-of-the-art[14]	Mature	Mature	Prototype	Concept	Mature	Prototype	Concept	Prototype	Mature
Conversion efficiency (%)[15]	50	60	85	40	60	35	85	85	N/A
Siting[16]	Inflexible	Inflexible	Flexible	Inflexible	Unrestricted	Flexible	Unrestricted	Flexible	Inflexible
Lifetime (years)[17]	30	30	20	20	5	10–20	30	10–20	30
Environmental impact[18]	high	medium	low	high	medium	low	medium	medium	high
Safety[19]	Exclusion area	Negligible danger	Containment	Low danger	Chem. disposal	Chem. handling	High voltage	Magnetic field	Radiation leakage

[1] PHES = Pumped Hydro Energy Storage
[2] CAES = Compressed Air Energy Storage
[3] FES = Flywheel Energy Storage
[4] THES = Thermal Energy Storage
[5] ECES = Electro-Chemical Energy Storage
[6] HES = Hydrogen Energy Storage
[7] CES = Capacitor Energy Storage
[8] SMES = Superconducting Magnetic Energy Storage
[9] NBL = Nuclear Base Load
[10] Power rating: This figure provides a summary of the currently available published data on power output for the best commercial and prototype installations of each system type.
[11] Reported capacity: In this row I have tried to provide an estimate of the storage capacity of each system type aggregated for known systems (e.g., PHES), and possible prototypes as indicated by the scientific literature (e.g. FES, CES, SMES).
[12] Response time: The time to respond to a demand surge is compared for each system type. Less than a cycle implies that the response will be instantaneous to all intents and purposes.
[13] Recharge time: The rate at which a given storage system can be replenished depends on many factors. The figures here are for a typical storage facility of 500 MW-h, assuming input power is not limited. Generally recharge times are similar to discharge times.
[14] State-of-the-art: Here 'mature' implies that commercial operation is well established, 'prototype' indicates that research is quite advanced, and 'concept' means that the research is at an early stage.
[15] Conversion efficiency: This is the 'round-trip' efficiency in transforming electrical power to stored energy and back to electrical power to the 'grid'. It does not include primary power generation or 'grid' losses.
[16] Siting: This row of the table provides a comparison of likely installation and construction difficulties.
[17] Lifetime: The estimates are crude but generally technologically complex systems have lower lifetimes.
[18] Environmental impact: Here impact means mainly visual degradation with buried systems having lower impact than storage in warehouse like buildings. Materials used in construction, and whether or not they have to be mined are included, and chemical disposal, including nuclear waste, is also taken into account.
[19] Safety: Apart from nuclear power, safety issues are relatively minor for these systems but not insignificant. This row gives a brief comparison.

Chapter 5
Known Knowns and the Unknown

A very Faustian choice is upon us: whether to accept our corrosive and risky behaviour as the unavoidable price of population and economic growth, or to take stock of ourselves and search for a new environmental ethic.

E.O. Wilson

An inefficient virus kills its host. A clever virus stays with it.

James Lovelock

I believe that a scientist looking at non-scientific problems is just as dumb as the next guy.

Richard P. Feynman

5.1 Diverging Supply and Demand

If one could imagine, and this is probably quite easy for many people, that 'business-as-usual' were possible to the end of the century, and that population numbers were to plateau at 10.5 billion, as is generally predicted, then consumption trends [1] suggest that we will require to find 25–30 TW of power in 2050 to feed the seemingly 'unquenchable thirst for energy' of industrialised, and modernising, societies. The trend is shown in Fig. 5.1, where the uppermost curve (solid line + diamond markers) depicts estimated power consumption for an energy profligate business-as-usual (BAU) scenario, while the lower curve (solid line + square markers) presumes that a slightly diminished rate of growth could occur due to 'peak oil'; i.e., rising energy costs owing to diminishing liquid fossil fuel reserves after 2020. In energy terms 25 TW for a year equates to 788×10^{18} J, or 747×10^{15} BTU. Ensuing calculations will be based on the less harmful 25 TW figure. Of this, half will be expended by industry, a quarter by transport, a sixth by domestic users and a twelfth by commerce [1]. But by 2050, if we have somehow managed, as we must, to wean ourselves off fossil fuels, 25 TW will not be there to consume, because renewables can provide only a frac-

A.J. Sangster, *Energy for a Warming World*,

tion of this. In Chap. 3 it has been demonstrated that as far as electrical power goes, the most that mankind can plausibly expect to extract from renewables is in the region of 14 TW, backed up by possibly ~2 TW from nuclear fission reactors. It is assumed that no major new energy sources become available, such as nuclear fusion power, deep sea wind power, or deep ocean wave power. Even if an engineering breakthrough were to happen in the next few years, the emergent technology is not going to progress through research, prototyping, commercial development and commissioning phases, quickly enough to impact on the energy mix by 2050. Consequently, once it is deemed sensible that fossil fuels should be left buried in the ground to curtail greenhouse gas emissions, mankind will experience a quite significant supply shortfall. Only 57% (14/25 × 100) of global demand for energy will be capable of being met from renewables. The shortfall could, in fact, be considerably more than this, given the inevitable unreliability of complex man-made systems, which the ecogrid surely would be, particularly if it is exposed to increasingly severe weather, possibly causing regular localised breakdowns. The possibility of sabotage, and worse, by humans genetically disposed to conflict, also has to be taken into account. Perhaps 30% would be a more realistic estimate of the power deficit. This shortfall is so large that even if mankind were to abandon all environmental degradation and safety concerns, in order to cover significantly more of the planet with renewable power collecting farms than is deemed prudent in Chap. 3, the difference is unlikely to be bridged. In any case, would the resultant much degraded planet, represent a desirable place to live?

Clearly therefore, in the post fossil fuel era, BAU is not an option, simply on the basis of two fundamental technology constraints, both of which have been addressed in this book. First, there are severe geographical, geological, and engineering limitations on the extent to which renewable resources can be exploited, thus placing a cap on available energy. Second, there are inherent reliability difficulties associated with a complex electricity supply system of global reach. Neither the degree of complexity, nor the proposed reach, is negotiable, if reasonably dependable production and transmission are to be secured from intermittent sources. Some small countries like Scotland, which is swept with the Atlantic 'trade winds', is blessed with a long coastline bordering a restless ocean, and possesses a plethora of islands that generate strong tidal flows, might believe that it is possible to 'go it alone' in converting to renewables. The big obstacle to doing this is intermittency of supply. Scotland, since its land area is small making diversification difficult, would have to massively overinvest in storage facilities to build and maintain huge reserves capable of smoothing out the peaks and troughs of supply. It is more efficient, more effective, and ultimately more sustainable to be part of a large continent wide system, linked into a global system. Nowhere, not even Scotland, can be an 'island' of self-sufficiency in renewable energy terms.

Where we are now, and where we have to get to is clear; the challenge for others is to devise a strategy for getting from a decaying and dysfunctional fossil fuel based world to a less energy profligate, sustainable future, powered by the postulated 'ecogrid'.

It is evident that at a global level, demand for power will have to be moderated downwards as the century progresses, so that it comes below the impending ceiling on power supply, which will inevitably manifest itself as we edge towards total reliance on renewables. So how can this moderation process be brought about? Available statistics indicate that global power consumption [1] in 2003 was just over 14 TW (Fig. 5.1), with the population at 6.7 billion. This suggests that we need to return to the 2003 levels of consumption by the time the ecogrid system, with some nuclear back-up, is in place and fully operational, whereupon parity between supply (assuming an optimistic 90% delivery) and demand will fortuitously be achieved. Unfortunately, this outcome would entail a global holiday for economic growth, for the next 40 years, and the enactment of policies to place a cap on global population. Politically, both of these measures are utterly 'off the radar', so it is pertinent to consider whether or not less unpalatable engineering solutions exist.

The demand for energy in human societies, as we have seen, falls into four main categories. In 2050, industrial, commercial, and domestic consumption are predicted to absorb 75% of the total power generated, while transport will employ the remaining 25% [1]. Power consumption for the non-transport sector is shown as a solid curve with triangular markers. At the end of Chap. 2 it was demonstrated that, of the energy supplied to electricity power stations in the form of fossil fuel or other energy sources, only 10% of it is actually used productively by the consumer. The rest dissipates as joule heating in generators, transformers, transmission lines, and in user equipment and appliances. These inefficiencies in the 'electrical sector' of the economy are undoubtedly paralleled in other areas where energy is expended. But it is the electrical savings that are important since the future is 'all electric'. Clearly considerable savings are distinctly possible, but it would require concerted action to improve efficiency in all areas of power usage, such as heating, lighting, manufacturing equipment, farming equipment, power tools, electrical appliances, gas appliances, computers, office equipment, electronic devices, etc., to make it happen.

It is surprisingly difficult to elicit relevant and helpful statistics from the literature, in order to form an accurate estimate of the energy savings, which could be made in the non-transport sectors. Perhaps, some clue as to what is possible can be surmised be comparing Switzerland and the USA, two industrialised nations with apparently similar gross domestic products (GDP). On a per capita basis, Switzerland has been shown to use only 20% of the energy expended by the USA [2], to achieve a similar standard of living. Furthermore, some studies suggest that the domestic sector in many parts of the world could be 80% more efficient than it is now [3, 4]. Globally, therefore, it is not unrealistic to anticipate that coordinated and strenuous efforts at efficiency improvements, year-on-year, could reduce energy consumption by at least 50% by mid-century. This would require that the growth rate in the consumption of power in the non-transport sectors should fall, from ~1.5%/year to zero, simply by enforcing stringent efficiency standards or by making inefficiency very expensive. However, despite such savings, these sectors will still need to consume ~14 TW by 2050, to continue day-to-day activities in support of a growing population all seeking 'western' lifestyles. As we now know, this is equivalent to the total power that can be extracted from renewables, by means of a fully developed ecogrid system operating at about 90% of capacity. Of course even 90% of capacity will probably seldom be available simply because of maintenance and replacement requirements and we should aim to maintain economic activity at a level which keeps energy demand well within the capacity of the system. Remember that supply intermittency has already been built into the available power estimates.

It is possible that mankind will be encouraged or persuaded to turn to the nuclear supply industry, rather than make sacrifices – but this will be a rather pointless short term option which will hardly 'ease the pain' as we have already observed. By building a nuclear power station a week until 2050 it has been suggested [5] that global nuclear capacity could possibly be expanded to 1 TW. Unfortunately at this rate of build, readily accessible reserves of uranium run out at about 2040 [6]. Of course liquid metal breeder reactors with their potentially dangerous plutonium legacy could be contemplated as a possible 'fix'. But in referring to their low breeding ratio, and very poor economic prospects, the renowned scientist Edward Teller, a staunch advocate of all things nuclear, is quoted as saying: 'Breeders don't work'. This still seems to apply, even today, although evidence for progress is growing! Breeder reactors in various guises are being contemplated, such as integral fast reactors, and thorium reactors, but none (as of 2008) is close to commercialisation. Consequently, it seems fair to assume that this technology will be largely irrelevant to the problem that faces us of achieving a massive growth in clean electrical generation capacity, over the next twenty to thirty years. The nuclear option, which would have to rely on the current generation of fission reactors, has a potentially limited future. However, as we have postulated in the previous chapter, it can provide a reliable source of useful base load for the proposed eco-grid, during the transition process when effective storage systems are being developed and commissioned.

Real and substantial energy savings are possible, and many of these could probably be enforced by introducing a marginal energy pricing system, in which base-load electricity and gas/oil for essential requirements would be easily affordable, whereas for consumption demands beyond the base level the price/joule would rise very steeply, so attracting increasingly expensive bills. How to do this at a global level is outside my area of expertise, and others with appropriate knowledge and skills will be required to devise a workable procedure. However, hopefully we would see disappear, many uses of energy, especially in modern industrialised societies, that are frankly trivial and unnecessary. There are lots of examples, in the home, in entertainment venues, in the gymnasium, in the garden, in the workplace and elsewhere. At the time of writing, on one of the few days this summer, in the south-east of Scotland, when the rain has stayed away and the sun has made a welcome appearance, the pleasure of decamping to the garden has been spoilt by noise pollution. The culprits are, of course, lawn mowers (mainly electric but petrol driven version are also a pest), but today there is also an electric hedge trimmer grinding in the background. For able bodied human beings why are such devices necessary? Much less noisy push-mowers, and hand operated hedge shears, were more than adequate to maintain the trim appearance of out-of-doors suburbia in the not too distant past. The manual versions also provided superb exercise for the user – surely a consideration in these days of spreading obesity?

Given that on average, during a working day, an adult human being is capable of providing muscle power of the order of 250 W [7] it is salutary to note that beyond 2050, a conservative 3 billion or so adult, able bodied, men and women (~30% of the total population) on the planet, will represent available power for doing mechanical work of 0.75 TW. If all of this muscle power could be used to do work that is currently being done by hand tools and other machines designed to boost human indolence, all powered by electricity, and remembering that electricity generation and transmission is, at best, 50% efficient, 1.5 TW (~10%) of renewable power generation could immediately be saved. This is the output of about 1500 large power stations! With so much muscle power at our disposal, why do so many trucks, delivery lorries, removal vans, garbage collection vehicles seem to have hoists or cranes using the power of the engine to lift goods on and off said vehicle, rather than use man power? The answer, of course, is easy access to ridiculously cheap fossil fuel energy. But in a resources strapped world, an awful lot of scarce energy can be saved by re-introducing muscle power. It is not so long ago, certainly within the memory span of anyone over 50, that coal delivery men were nonchalantly shifting 1 cwt coal bags on and off trucks using their own 'brute strength'. The construction industry has also got rid of the 'muscle power' and the manual techniques that were more than good enough, in the not too distant past, to create the sophisticated buildings and structures appropriate to the needs of societies that were well advanced even by today's standards. Others are quite free to contemplate the further savings that could be procured by making intelligent and imaginative use of the muscle power of horses, elephants, yaks, or oxen! Of course health and safety, and animal rights, issues would have to be addressed, but the rules may possibly change when energy is in increasingly short supply. It is, perhaps, pertinent to emphasise, that we are contemplating here the restoration of the health and safety of the planet itself, so it seems inevitable that unpalatable choices will have to be made at some stage!

Less controversially, savings can undoubtedly be procured by introducing clockwork, solar cell, and perhaps kinetic mechanisms, into toys and electrical and electronic devices. Many free standing electronic devices are increasingly being supplied with solar panels to power the electronics – such as calculators and watches. This could be extended to a much wider range of electrical components, as solar cells become more efficient, and more robust. Apparently a 40 W solar panel has recently been fitted to a hopefully quiet lawn mower [8], a clear indication that this technology has reached a stage where it is justifiable to suggest that significant savings in electricity usage globally, could soon be procured without seriously encroaching on individual liberties. My guess is that a further 20–30% saving in energy usage could be achieved, post 2050, by well directed and focused efficiency programmes, aimed at suppressing the worldwide manufacture of frivolous, mainly electrical gadgets, but also other unessential powered products. The

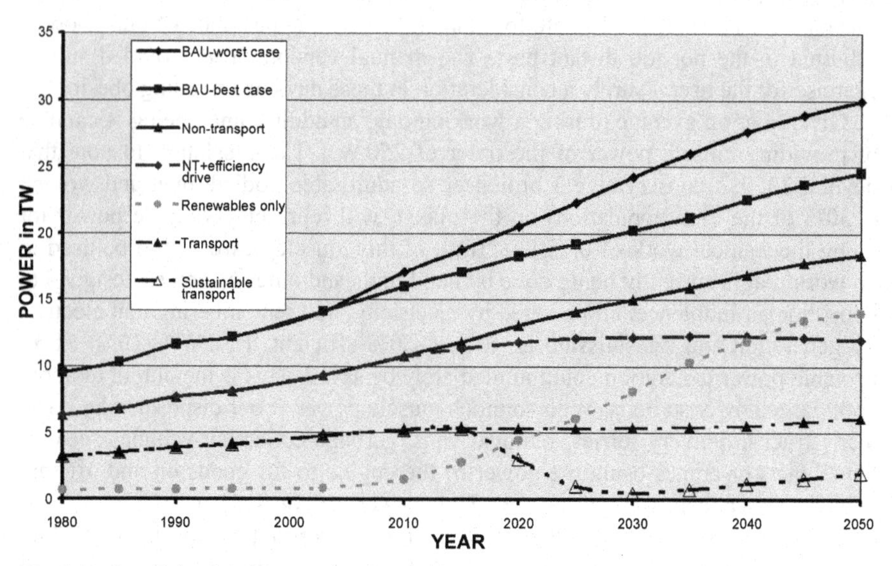

Fig. 5.1 Growth in global power consumption in terawatts between 1980 and 2050

object must be to increasingly introduce manual, solar-powered and clockwork powered devices and appliances into the market. Savings of the order of 25% have been predicted for such programmes in a recent report from the McKinsey Global Institute [9]. If all these savings could be implemented, the non-transport sector would be seeking to consume 12 TW, or about 85% of the available power from renewables, towards the second half of the century, assuming the full capability of renewable power sources has been brought on stream – a big assumption. In Fig. 5.1, the way in which non-transport power consumption could diminish, if the kind of savings outlined above were to be implemented, is represented by the dashed curve with diamond shaped markers. It can be seen that power consumption for these sectors falls below the available power from renewables plus nuclear base load (assumed to be operating at 90% full capacity: dotted curve/circular markers: see Sect. 5.2) at about 2040. The big question is: can transport be accommodated within the remaining 15%? What kind of transport infrastructure can be furnished when it is capped at a power level of about 2 TW?

5.2 The Transport Crunch

As our freedom to burn fossil fuels becomes increasingly constrained by critical levels of CO_2 in the atmosphere, travel by road and air, in vehicles and aircraft that are wholly dependent on these fuels, represents an activity, which eventually and unavoidably, will be possible no longer, in its present form. It is taken for granted, in saying this, that with a global population in excess of 10 billion the use of land area for bio-fuels is unlikely to be tolerated by hard pressed humanity. Self-evidently there will be an expanding need for food production, and since this may become increasingly difficult to achieve within an unpredictable and less benign bio-sphere, productive farmland will be too valuable to be given over to bio-fuel crops. Also, it is not unreasonable to anticipate that there could well be considerable pressure to recoup land in order to re-introduce natural forests as it becomes important to replenish some of the planet's CO_2 sinks. Consequently, without fossil fuels or bio-fuels, the provision of transport systems, which in any way resemble what we have now, is likely to become one of the most difficult challenges faced by engineers in the second half of this century.

The BAU trends [1] suggest, as indicated earlier, that transport will consume 25% of future energy needs. Thus, with total BAU demand predicted to rise to at least 25 TW beyond 2050, maintaining a transport infrastructure with the capability, capacity, and versatility of the arrangements that we currently enjoy, would obviously involve consuming just over 6 TW by 2050 (see chain-dashed curve with solid triangle markers – Fig. 5.1). Flying by jet, and travelling by car, are activities that employ power essentially to overcome gravitational, inertial and frictional forces, and for any given vehicle this power is relatively independent of how it is generated. Consequently, given that fossil fuels represent the most efficient way of energising vehicles, it is virtually impossible for transport consumption to fall much below 6 TW in any future BAU scenario. This figure is already much more than seems likely to be available from renewables (about 2 TW), given rising population pressures for land and growing environmental issues. However, there are plenty of technological proselytizers, who would claim otherwise, and it therefore seems prudent to examine the evidence. Let us, for the sake of argument, assume that a BAU transport scenario could be pursued in the post fossil fuel era; in which case what electricity based technical solutions are available to do so, and what is the demand for energy likely to be from this sector? Given that electrical supply inefficiencies are reasonably well known, it should be possible to answer this question by calculating all the incremental power losses, associated with energising vehicles from electricity.

There are two favoured vehicle propulsion modes for a post fossil fuel world [10]. These are hydrogen and the electrochemical battery. It can also be stated that the high consumption elements of the transport sector are air travel and private car use. For air travel the only possible replacement fuel, when oil, from an ecological perspective, is deemed much too harmful to use in combustion processes, is hydrogen. It is plentiful enough in theory, and energetic enough in practice, to power large commercial aircraft. Hydrogen powered aircraft have been subject to many studies since as far back as 1980, and it has been suggested that liquid hydrogen will first be used in a large aircraft [2] such as a Boeing 747. An aircraft of this size would allow liquid hydrogen to be stored in the fuselage as well as in the wings, for example in the upper first class compartment. The extra storage space is required because, while hydrogen exhibits a slightly superior energy density per kilogram than kerosene, it is obviously much lighter than jet-fuel, and consequently a much larger volume of the aircraft has to be set aside to carry enough liquid hydrogen (typically 45,000 kg or 646,000 L) to permit the aircraft to function to 'modern' standards in trans-continental roles. Passengers in such aircraft would literally be encased in liquid hydrogen, in storage tanks above their heads and below their seats. Given acute public knowledge of the Hindenburg disaster in May 1937, it seems valid to question whether or not travellers of the future would be willing to take to the air enclosed in an oversize cigar shaped pod with a liquid hydrogen 'skin' cooled to a sub-glacial –253°C.

Despite doubts about the practicality of these aircraft, it is informative to examine the energy requirements that will be needed to provide BAU levels of air travel in hydrogen fuelled airliners. Of the predicted 6 TW of power consumption associated with the transport sector as a whole by 2050, about one-sixth can be attributed to mass air travel [11], if predicted trends are believed. In a world replete with fossil fuels this would amount to 1 TW produced by burning kerosene in millions of jet engines per year. Hydrogen has an energy content of 2.3 kW-h/L, and since 1 TW for a year equals 8.8×10^{12} kW-h, we can deduce that air travel based on hydrogen powered aircraft will require 3.8×10^{12} L of the gas. Actually this is probably an under-estimate since wider bodied hydrogen jets will suffer about 28% more drag than current aircraft [2]. On the other hand, H_2 fuelled aircraft can fly higher than current kerosene powered jets so some lowering of drag can be allowed. It seems reasonably valid to suggest that a 20% increase in fuel requirements for H_2 powered air travel could apply, giving us a figure of 4.6×10^{12} L. The electrolysis of water to generate H_2 requires 3.5 kW-h/L, as we have seen. This means that the electrical power needed to generate sufficient hydrogen per year to support this level of air travel is 1.8 TW. Additionally the hydrogen has to be liquefied and this also takes power. A figure of 12.5–15 kW-h/kg or 0.87–1.05 kW-h/L applies to the liquefying process [10], so a further 0.5 TW is needed to produce liquid H_2. A total of about 2.3 TW of electrical power will be required to maintain air travel at BAU levels in the post fossil fuel age. Some of this, perhaps 10–15% could be attributed to the non-transport sector to represent the energy costs of mining and refining fossil fuels, but this still leaves a consumption level that is impossible to accommodate in any scenario of the future, in which renewable power is capped at about 14 TW.

Post the fossil fuel era, most future predictions envisage that road vehicles, apart from trams and trolley buses, will be propelled by means of a hydrogen fuel cell, or by means of a rechargeable battery. In both cases electricity, generated using renewable sources of energy, would be used either to produce hydrogen by electrolysing water, or to provide vehicle battery recharging. Both of these processes are inefficient – 70% for electrolysis, as we have just seen, and 60% for battery charging. For road vehicles it is usually recommended [10] that compressed hydrogen gas, rather than liquefied gas, is employed largely because hydrogen is liquid only below the rather numbingly frigid temperature of –250°C. Fitting refrigeration systems and cryogenic storage tanks in cars that can maintain these kinds of temperatures is highly impractical and pressurisation (at typically 3600 psi [10]) is usually recommended. However, hydrogen storage at high pressure incurs significant additional losses, since compressors are only about 60% efficient. Consequently, the efficiency of hydrogen production for vehicle use is, at best no more than 40%. The net result is that the extrapolated trends, which predict that by 2050 the consumption of fossil fuel by road vehicles will rise to an equivalent power consumption level of at least 4 TW, point to a massive 10 TW (4 TW/0.4) being demanded from the renewable electricity supply system to produce the required hydrogen. To this should be added all the energy costs associated with setting up a network of hydrogen stations, analogous to petrol and diesel oil stations, and the energy expended in servicing these stations. The capped supply and the predicted demand are now completely irreconcilable! To travel as we do now we would have to give up all other uses of energy! It seems appropriate here to quote from *The Hype about Hydrogen*. In it, Joseph Romm [10] is motivated to comment that it hardly makes 'much sense to generate electricity from renewable resources, then generate hydrogen from that electricity using an expensive and energy-intensive electrolyser, compress and liquefy it (using more energy) ship the hydrogen over long distances (consuming more energy), and then use that hydrogen to generate electricity again with low temperature fuel cells in cars'. On all the available evidence it is hard to disagree.

The unavoidable conclusion is that cultures that embrace private cars, road transport, and cheap air travel – obviously a strong feature of the present day industrialised world – are quite incompatible with a predicted energy capped post fossil fuel era. Once the populace of the globe comes to realise that private cars, long distance road transport, and air travel are impossible to sustain as the oil supply is throttled back – that there is no 'silver bullet' in the form of hydrogen – it seems likely that between 2015 and 2050 the skies will become devoid of vapour trails and the motorways will become a haven for cyclists. This will be hugely beneficial to the health of the planet. A possible, but perhaps rather too optimistic, representation of this trend is shown as a chain-dashed curve with unfilled triangular markers in Fig. 5.1. It has been inserted purely as an illustration of the potential impact on transport of the coming decline in oil. It is merely one of many possible power/energy allocation scenarios depicting the transition to a post fossil fuel future. Of course, the sooner the break with the era of cheap petroleum begins the greater will be the benefit to the planet in reduced carbon emissions – but responses so far, to the global warming threat, suggest that self-indulgent human beings will inevitably 'drag their feet'. Today greenhouse gas emissions associated with the transport sector [12] are at about 13% of the total, and they are rising rapidly. On current trends, transport will contribute about 2 billion metric tons equivalent (2×10^{15} gC/year) of greenhouse gases by 2015, and this could fall to less than 0.2 billion metric tons by 2030, with a hopefully accelerated flight from fossil fuels. This change would produce an emissions reduction that is about a third of the total emitted in 2008. Needless to say, it would not be too concerning, if the process depicted on the graph were to be delayed a little, because mankind chose to direct significant levels of industrial and engineering effort into the construction of a competent version of the ecogrid, which would, of course, involve some fossil fuel burning to provide the required energy for the building process. Actually, the expedient route forward may demand a relatively slow phasing out of fossil fuels because of the powerful influence on climate of aerosols, which can range from the dust ejected by volcanoes to the particles emanating from smokestacks and vehicle exhausts. Scientists now believe that aerosols have a cooling effect on the atmosphere, and consequently that it could be unwise to allow them to clear from the atmosphere too quickly.

Despite the loss of the 'products' of the automobile and aeronautic industries, *mankind* will not be reduced to *manpower* to get about. A power budget for transport of 2 TW is very considerable (roughly what the transport sector burned in 1980), and will allow a significant level of power assisted travel, but it will be largely in the form of ground based mass people-movers, i.e., trains, ships, trams, and trolley buses. Without dipping too much into the area of future prediction, which is not a skill usually possessed by engineers, it seems appropriate here to try to make some extrapolations based on well established technological trends. Hopefully by doing so, we can gain some understanding of what the major developments in the transport sector might be when all energy comes in electrical form from renewables, and when, more importantly, it is severely capped.

The biggest development, it is reasonably safe to say, will be in electrified railways. There will be much more of them serving a much wider community. The expansion of the railway system will become a high priority for governments once flying becomes no longer affordable, particularly high speed international, and transcontinental, systems. The rapid expansion of such systems may well take advantage of the emptying and freeing up of motorways as road traffic dwindles owing to the high cost or unavailability of fuel. Converting motorways to high speed railways will be much easier than developing new networks. A power budget of 2 TW will accommodate an awful lot of rail journeys. While this form of travel will be the primary replacement for air travel for those that have to journey long distances relatively quickly, it is also easy to see that much of the need for roaming around the globe that has been considered necessary in the past, is already being undermined by the massively improving accessibility of wideband communication systems and the internet, through the agency of high speed digital electronics. For example, electronic conferencing for large groups of people scattered around the globe will become commonplace, eliminating one incentive to travel for large numbers of individuals. In 30–40 years electronic communications systems will be much more sophisticated than they are now, with computer processors continuing to increase in speed and memory capacity and broadband high speed interconnections getting faster and more reliable. Clearly, many of the reasons for travel that existed in the past are being eroded.

At the beginning of *Heat* [13], the author recounts a revealing incident at a time when he was still evolving his stance on global warming. Following an oral presentation he was asked a question to which he recalls being stumped to find an answer. It was at a seminar in London in 2005, which had been convened to address the problem of greenhouse gas emissions and the need for an 80% reduction. The question was: 'When you get your 80% cut, what will this country look like?'

He recalls referring the questioner to Mayer Hillman [14] who happened to be in the audience, and Hillman's brief answer was 'A very poor third world country'.

One's first reaction to this statement is to suggest that perhaps there will be no first or third world countries once the polarising and divisive 'black-stuff' is left buried in the ground where it is safe? As an electrical engineer and scientist, my second is that it is unduly pessimistic and takes no account of the fact that the problem is global. Human beings have come a long way in their development of science and technology, and they will certainly continue to be innovative, if they are allowed to be. In other words, if the planet remains habitable, because we manage to avoid inducing run-away warming, our much less mobile human societies will probably embrace the 'virtual world'. This world will be made possible by the increasing availability of low power consumption, highly efficient electronic devices, and powerful computer systems, married to ultra-fast, ultra-wideband communication techniques, as has been suggested in earlier discussions.

In a recent article in *The Herald* of the 6th December 2007, the nascent possibility of virtual systems is clearly illustrated. The article accompanies a picture of the Colvilles Pavilion and Tait Tower built for the Glasgow Empire Exhibition in 1938, and demolished some 70 years ago. Entitled '3D images bring 1938 Empire Exhibition to life' the story describes a digital recreation of the exhibition which was opened to the public on the 5th December at Bellahouston Park, a park on the south side of Glasgow, where the original exhibition was sited. Visitors were able to explore the 430 acres of the exhibition using the 3D technology, which gave the individual the vivid experience of walking among the original buildings. The project's creators, working from archived architects' drawings, sketches and photographs of the exhibition, used three dimensional visualisation and digital imaging to achieve a realistic representation of every building. The article reports on the reaction of Percy Walker, aged 91, who had visited the original exhibition. He was quoted as saying 'The digital recreation is absolutely amazing. The only thing missing is the people walking about'.

Computer programs, which professional designers can use to present their designs in customer friendly, interactive, three dimensional modes are becoming common place. Such programs are now available for a wide variety of design disciplines with interests ranging from buildings, to aircraft, to automobiles, to clothes. For example, in architecture it is possible using the latest imaging techniques to provide clients, the general public, and/or interested viewers, with very realistic 3D presentations of the exterior and interior spaces of new designs. Armed with suitable equipment these will allow the viewer to interact visually, aurally and tactilely, with the proposed structure. The best of these virtual systems can lull the viewer into believing that they are actually wandering through, and interacting with, the real 'building' before even a 'sod has been cut'. The possibilities for this type of technology are manifold. A research team at the University of York [15] in the UK is well advanced in the development of a virtual reality experience which they term 'Virtual Cocoon'. The wearer is lulled into believing that he/she is on safari in Africa – they can see it, feel it and smell it. One of the team has been quoted as saying: 'For me the project will be finished when someone puts the helmet on and they don't know whether what they are experiencing is with or without the helmet on'.

With apologies in advance for the following smidgeon of fanciful speculation, but it seems well within the bounds of technology to replace the Empire Exhibition, or a designer town, with the architectural jewels of Venice, Florence, Rome, St Petersburg, San Francisco, the Taj Mahal, Machu Picchu, the Pyramids, etc., all accurately and sensitively imaged in glorious three dimensional detail, probably using holographic techniques. Then perhaps we have a means of weaning people away from what has become a highly damaging tourism industry, largely because of the vast numbers now participating, all 'riding on the back' of ridiculously cheap air fares. It is admittedly difficult to see how one could persuade airport hopping junkies to replace their weekend trip to Prague (or any capital city with a worn out historic centre) with a virtual reality headset experience. But perhaps there is some room for optimism, insofar as many reputedly never really see the over-populated cities or countries they go to visit, because they are side-tracked into partaking of 'entertainments' they could have obtained nearer home, or are put off by the crowds and the pollution, or are discouraged by pick-pockets, beggars and con-merchants, or stay for their safety within secure walled and fenced holiday hotels and condominiums? Artificial sunshine, artificial suntans, and artificial beaches are, of course easy. All 'holiday needs' could be available, in the future, at a local emporium, and all powered by renewable electricity!

At a local and regional level it is apparent that renewable technology will favour trams and trolley buses for mass travel within cities and towns, probably backed up by battery powered buses in less populated areas. It also seems obvious that since fuel cell vehicles that operate on hydrogen made from electrolysis use four times as much electricity per mile as similar size battery electric vehicles [10], hydrogen power will not provide a significant part of the mix. It is not unreasonable to predict that local buses and local delivery vehicles will be solar power assisted (at least during daylight hours) to enhance range and flexibility. In towns and cities battery powered taxis, again boosted by solar power will be the norm. The pioneering work by Sir Clive Sinclair, back in 1985, with his ill fated, battery powered, three wheeler car the Sinclair C5, which was apparently solarised in 1990, gives plenty of reason to believe that battery powered, and solar assisted, compact 'city cars', and power assisted bicycles, could be a feature of post 2050 townscapes [16].

Airship developments geared to the post fossil fuel age are already being planned [17] as a cursory visit to the worldwide web will soon demonstrate. These, mostly helium filled aerodynamically shaped craft of the future are heavier than air to improve controllability, gaining lift from vectored thrust from their engines, from volume adjustment of the helium container, and in forward flight from aerodynamic forces. Such ships will be three times faster than surface ships. However, the use of helium, while safe, has sustainability problems. It is produced from natural gas, which has a limited lifespan, and helium, once released into the atmosphere, is unrecoverable since it escapes into space, unlike relatively heavy carbon dioxide molecules. Hydrogen, of course is the answer, but it is potentially very expensive to produce, and because of the Hindenburg incident there could be a serious public resistance issue with H_2 airships. Nevertheless, in a recent patent [18], protection is sought for the idea of:

> A lighter-than-air ship using hydrogen or other gas as a lift gas with at least one hydrogen fuel cell aboard. The fuel cell can draw hydrogen fuel from the lift gas reservoir to produce electricity both for the ship's use and optionally for propulsion. The waste product of the fuel cell is water which can be used for the needs of a crew on the ship. The hydrogen lift gas chamber, which can be compartmentalized for lift control, can be surrounded by a safety jacket filled with an inert gas and contain optional hydrogen and/or oxygen sensors.

Gas envelopes covered in thin film solar cells are also being mooted to provide electrical power to silent motors. Solar cells can, of course, only provide electric power during the day, and consequently they cannot be the sole source of power. At present, hydrogen provides the obvious back-up, with perhaps seaweed generated bio-fuel appearing in the longer term. Given that airship technology is clearly in its infancy, it is not too difficult to envisage safe, sustainable and environmentally friendly, cargo and passenger vessels being developed well before the demise of the fossil fuel age.

Hydrogen is also the fuel of choice for powering environmentally friendly surface ships [19], but because the cost of production may be prohibitive, for reasons discussed earlier, other solutions for sea and ocean travel may be preferred, in the long run. Concept vessels, which employ a range of energy sources, aimed at eliminating greenhouse gas emissions, have been introduced to the world over the past few years, essentially to demonstrate what is possible with known technology. For example, a cargo ship designed to run exclusively on renewable energy was presented at World Expo in 2005, by the Scandinavian company Wallenius Wilhelmsen [20]. Self-sufficiency in energy is achieved by harnessing the power of the wind, the waves and the sun, backed up by hydrogen, and hydrogen fuel cells. The ship's design incorporates a cargo deck area equivalent to 14 football fields, and it is large enough to carry about 5000 containers. Three giant rigid sails manufactured from special lightweight composite materials are covered in flexible solar panels to help drive the ship at its cruising speed of 15 knots. Wave power is harnessed by means of a series of 12 fins on the underside of the pentamaran hull, and the fins transform the wave energy into mechanical energy, electricity, and hydrogen. The fins double as propulsion units, driven by electric motors using electricity generated from the onboard renewable energy sources. This propulsion system eliminates the traditional stern propeller and rudder arrangement.

The technology to provide sustainable, environmentally friendly, and effective transport systems, once the fossil fuel age becomes history in perhaps 40 or 50 years, is already in existence – or if not, it is sufficiently close to commercial realisation to enable one to confidently assert that worldwide travel in the second half of this century, although perhaps slower than we are familiar with today, could be comprehensive, substantial and far reaching.

5.3 Towards a Wired World

There is probably a multitude of possible routes that governments could follow in attempting to hasten the transition to renewable energy. From a rational standpoint, if maintaining modern standards of living is the goal, it seems inescapable that their obligation is to do everything possible to achieve sustainability. In the event that politicians and governments are brave and wise enough to make the huge commitment that is called for, their deliberations about how they may intervene to safeguard the future for generations to come, should be founded on the guiding principle, enunciated below. It would surely be morally indefensible for decision makers to do other than heed it. The principle is contained in the following quotation [3], which proposes that they should:

> focus on the ethics of what we reasonably may, and may not, do to the future. We may reasonably impose on future generations some reduction in their enjoyment of industrial and technological progress; we may not, however, destroy the irreplaceable essentials of the biosphere – such as healthy soils, water, atmosphere, wildlife, forests and other ecosystems, and equable climate. Thus we should make our main objective the prevention of serious climate accidents and the ruinous catastrophe of runaway global heating, whose essentially limitless costs would overwhelm all the benefits of economic growth and human progress.

Furthermore, in formulating a 'road map' to the post fossil fuel age, it is difficult to believe that pressure groups and our leaders will not be influenced by the events of 2008, which have resulted in the worldwide collapse of banking, a shaking of the foundations of modern exploitative capitalism, and perhaps a shift to a new economic order in international relations. 'Every cloud has a silver lining', as the well known saying goes, and from the viewpoint of those interested in addressing the dangers of global heating, this crisis has had one unexpected benefit: it has forced all the world's governments to realise that they have major issues in common and that when the challenge is sufficiently threatening to their common wellbeing, they are capable of a high degree of cooperation to mobilise resources on a previously unimaginable scale. Building the international ecogrid will require cooperation in a similar vein. Within a matter of months of the credit crisis breaking, thousands of billions of dollars ($> \$ 10^{12}$) were poured into the banking systems of the industrialised world. It was a rescue programme of unprecedented size in the span of human history. It is difficult not to conclude that if it is possible for cooperating nations to bail out banks with such huge resources at such short notice, it is surely possible to believe that they can draw upon similar commitments at the global level, to make the equally important and critical 'bail out' from the fossil fuel age, by making funds available on a 'credit crunch' scale to facilitate the ushering in of a new era of renewable power generation and distribution. As has been wisely noted elsewhere [21]: 'this is the kind of purpose built economic direction that normally functions in wartime, and in periods of national emergency'. Instead of letting the market dictate the pace of change as we have done until now, with negligible evidence of progress, governments will have to decide coolly and rationally on the appropriate transition process, and then devise market structures to help deliver it.

Burning fossil fuels is now known to be foolhardy given the immense damage that greenhouse gases are reportedly doing to the ecological wellbeing of the planet. Oil and gas are also being depleted fast as economic growth and the spread of industrialised world lifestyles proceeds apace. Should we continue to spend this fossil fuel capital, or should we use it to build the ecogrid or its equivalent, and attempt to move humanity away from its dependence on fossil fuels by 2030, rather than 2050–2060, when oil is predicted to run out, and when one way or another we will have to lock all other fossil fuels below the ground for a very long time? By then the CO_2 levels in the atmosphere may have reached a level where burning fossil fuels, even to provide energy to build an ecogrid, may be simply self-defeating because of the real danger of triggering run-away global warming.

So what, in engineering terms, is the way forward? At the end of Chap. 4 it was noted that to achieve 14 TW of renewable power by 2050, we will need to build 52,000, 250 MW power plants at the rate of three a day for the next 41 years! With the exception of a relatively small number of new hydro-electric and geothermal power plants, on-site construction of the vast majority of renewable stations and storage facilities will generally require a moderate level of civil engineering and a high degree of assembly of factory built units, such as turbines, generators, transformers, control systems, switching systems, refrigeration systems, battery units, coils, flywheels, etc. Many millions of these factory-built units will be required in an undertaking, which in character and scale will actually not be unlike the manufacturing and assembling of cars, vans, lorries, buses and aircraft. Given that these affectations of modern man are immensely harmful to the planet, and given that with the dwindling availability of oil they will become scrap by around 2050 in any case, the engineering logic is clear. A newly formed, United Nations administered executive, empowered to implement the ecogrid, should decide that car and aircraft manufacturing be terminated, to make all of the fabrication and assembly plants of the automobile and aeronautic industries, and the vast number of suitably skilled and qualified engineers in these industries and their suppliers, available to contribute to a declared eco-war effort to build an ecogrid. The scrap from useless planes and road vehicles will help to provide the massive volume of materials, particularly metals, which the manufacturing of an ecogrid infrastructure will require. The 'executive' will have to be vested with the power to short-circuit or accelerate planning issues associated with the rapid and extensive development of renewable power stations around the globe, some inevitably in environmentally sensitive locations. It will also have to initiate a massive recruitment programme to encourage new generations of students into engineering and science. Perhaps the 'executive' could consider the introduction of targeted inducements, to attract suitable youngsters into renewable energy engineering courses in institutions around the world. It will be critical to the success of such a project, that the supply of suitably qualified engineers keeps pace with demand, up to 2050 and beyond.

In 2005 the automobile industry worldwide, produced 66.5 million vehicles, including cars, light commercial vehicles, and heavy commercial vehicles [22]. To manufacture these, the industry employed approximately 2 million professional engineers and skilled technicians – about 25% of the total workforce of 8.5 million. The aeronautics industry employed close to 0.5 million staff with again about 25% of these employees being engineers and technicians. So the automobile and aeronautics industries are capable of providing just over 2 million engineers to the ecogrid project. This does not include the myriad of small companies supplying components to the major manufacturers. One can only guess at the numbers of engineers employed in this sector – but if we engage now in an Enrico Fermi exercise in logic – we can perhaps suggest half as much again, about one million. Most of these will need to be employed in the development of wind and solar farm infrastructure, since wind and solar will provide by far the most significant increase in new renewable power capacity as Table 3.1 shows. The predicted installed power from these sources is 7.5 TW from wind and 4.5 TW from solar collectors. These renewable power stations will have to be sited in a wide range of locations around the world and are likely to be designed on the basis of providing installed power ranging from 10 MW to 1 GW, with an average capacity probably in the vicinity of 100 MW. This means that we will need close to 750,000 wind farms and 450,000 solar farms. As we saw in Chap. 3 wind turbines are capable of generating 3 MW of peak power on a good day, so a mean capability of 1 MW is a not unreasonable presumption. A simple division sum dictates that 7.5 million wind turbines will have to be produced and installed between now and 2050, and this in turn computes to 500 per day for the next 41 years. So continuing in Enrico Fermi mode, each turbine/generator set including aerodynamic blades, is, in automobile manufacturing terms, equivalent to building something like 8–10 trucks. Consequently, to manufacture the wind turbine parts will be equivalent to manufacturing 5000 vehicles per day. In the hypergrid scenario each wind farm will be backed-up by a MES facility capable of providing an equivalent power capacity for 8–10 hours. As we have seen in Chap. 4, storage systems vary hugely in form and complexity, from compressed air storage in vast caverns to magnetic energy storage in arrays of superconducting coils. In engineering terms the latter would be at least as technology intensive as the farm itself and will call for similar levels of manpower both in the manufacturing and assembly phases of construction. This effort is estimated to be equivalent to the production of 3000 vehicles per day. Consequently supplying the parts for the ecogrid power stations if these are being built at a rate of 500 per day will be equivalent to the automobile industry producing 8000 vehicles per day. A similar sum for solar farms employing (typically) 25 kW generator sets [23] will require the manufacture of 18 million by 2050. This equates to 1200 per day for 41 years. Each solar genset with its optical reflector and support structure, tracking control servomechanism, gear train, generator, sensors electronics, control electronics, safety mechanisms and protective enclosure will require effort equivalent to the manufacturing of 5–6 trucks, the solar units being smaller than wind farm equivalents. Thus solar farm components will be equivalent to the manufacture of

6000 vehicles per day. Adding to this the effort required to construct back-up storage facilities (3000 vehicles per day) gives a total for solar of an equivalent effort level of 9000 vehicles per day. In total for wind farm plus solar farm components we require a rate of production equivalent to about 17,000 vehicles per day. In 2005 the automobile industry managed $66.5 \times 10^6/365 = 182{,}189$ vehicles/day. Of course the vast majority of these were cars and light vans. If we gauge in manufacturing terms that a truck requires three times as many engineers as a car to assemble then at 60,000 trucks per day the current automobile industry is well capable of providing the components needed to equip the renewable energy farms required to supply the ecogrid by about 2050.

While the factories of the automobile and aeronautics industries are manufacturing the components, site construction and plant assembly will have to be proceeding in parallel. According to *Energy from the Desert* [24], it will take about 3000 man-years to construct a typical solar farm. Just to get a handle on the sort of man-power numbers involved here, if we assume a 40 year time scale for building the ecogrid, then 3000 man-years equates to 75 engineers and labourers per farm. As indicated above, a 14 TW ecogrid system will require 750,000 wind farms plus 450,000 solar farms, a total of 1,200,000. Therefore over 40 years we will require something like 90 million engineers and labourers to provide site construction man-power. In addition, construction of the grid system itself (pylon construction, pylon installation, low-loss cable development, cable stringing, very high voltage insulators, AC–DC conversion plants, up-converters, down-converters, etc.) will, at a guess, engage another 10 million engineers and labourers. In a planet supporting 6.7 billion people and rising, finding 100 million (1.5%) able bodied individuals peppered with enough qualified electrical, mechanical and civil engineers, from automobile, aeronautic and perhaps military sources, hardly seems likely to present a serious hurdle for the project, in quantitative terms. The political and economic implications of marshalling such a work-force are another matter!

The guesstimates are admittedly quite crude in the above, with facts and figures on man-power requirements in the renewables industry inevitably being sparse. Nevertheless, they are sufficiently representative to conclude that, when viewed in project engineering terms, building an ecogrid system by 2050 is certainly within the realms of possibility. The assumption that a major sector of the global industrial complex can be redirected towards the manufacture of renewable power station infrastructure is, self-evidently, the primary obstacle to implementation, simply because the political implications could be ‘too hot to handle’.

The huge cost and the massive effort required to embark on the construction of a global renewable energy supply system, which will secure a sustainable planet for future generations, probably entails moving away from market solutions and entertaining a more centrally, but democratically, directed economic model. A free market of global reach was not instrumental to the building of the Great Wall, to the building of the Suez Canal, to defeating the Third Reich, to getting Sputnik into space, to getting men to the moon. In *Slow Reckoning*, Athanasiou puts it this way:

> As awareness of biophysical limits increases it will become difficult to keep faith with small remedies. It is not impossible that soon ecological deterioration will routinely inspire echoes of William James's call for a moral equivalent of war [25], only this time as a war of cooperation, a war to save the Earth. That is what it will take [26].

If global warming, and the battle to counteract it, which calls for a 'gigantic shift' into renewables, is not an emergency, what is? Even a well known authoritative voice [27] in this debate has suggested, particularly in relation to coal, that:

> The most difficult task, phase-out over the next 20–25 years of coal use that does not capture CO_2, is Herculean, yet feasible when compared with the efforts that went in to World War II.

Furthermore, in relation to the long period of economic growth enjoyed by the US government on the back of fossil fuels, and the increasing resistance to the making of hard decisions, which this has promoted in US leaders, particularly vis-à-vis global warming and energy supply, it has been noted rather interestingly that:

> The pejorative part of the situation is the immurement of the body politic in the temporary comforts which come from spending capital. There must be an operative organ of cure. This is the appropriate political organization to see the body politic in possession of sufficient information to allow the capital investment to build equipment. What is needed is a National Energy Resources Executive (NERE). This would be an authority with *war-time* powers to gather capital and manpower, to organize its building programmes. [2]

This was written in relation to the US economic position in 1980, but if 'National' is replaced with 'Global' the observation is not inappropriate to the current international situation. The inference seems to be that scientist and technologists need to be much more pro-active in devising a route map to ecological health for the planet.

Normally the funding for the activities mentioned above would be raised from taxation. However, on a planet which is in danger of tipping into disastrous run-away heating a more appropriate solution to the funding problem is potentially provided by Tickell [3]. In Kyoto2 he proposes that on a yearly basis a global limit should be set for carbon pollution. Once the limit is decided, it becomes possible to compute the amount of gas, oil, and coal that can be burnt in any given year. Permits to burn this amount of fuel would then be sold to companies extracting or refining fossil fuels. All going well, these permits should filter down the supply chain to the end user or polluter, thus sanctioning his or her pollution. This has the advantage of regulating a few thousand corporations at the 'upstream' end of the fuel supply chain – rather than a few billion end users. These suppliers would purchase their permits in a global 'uniform price sealed-bid' auction, which would be subject to both a reserve price and a ceiling price, to ensure that the cost of permits does not harm the rest of the economy unduly. The auction, it is suggested, would be run by a coalition of the world's central banks. The funds raised would accrue to a Climate Change Fund to be invested in renewables and other activities by a UN appointed 'executive'. The demand for fossil fuels should fall, so that fewer permits will need to be issued in later years. In Kyoto2 the planned areas of expenditure for the carbon 'windfall' would be:

Clean energy research and deployment	$ 170 billion
Domestic energy conservation	$ 250 billion
Enforcement through national governments	$ 50 billion
Ecosystem maintenance	$ 300 billion
Adaption to climate change	$ 200 billion
Agricultural reform	$ 11 billion
Geo-engineering	$ 0.5 billion
Emergency relief and health expenditures	$ 12.5 billion

This totals to $994 billion, or almost $ 1 trillion, and represents the global bill to stabilise atmospheric greenhouse gases at 350 parts per million (carbon dioxide equivalent). $ 1 trillion is roughly 1.5% of the global economy in 2008. But to put it in perspective, at least three times this amount was lavished on propping up crumbling banks, by the world's major economies, during the 2008 'credit crunch'. What we can deduce from this, is that the expenditure levels that are likely to be required to seriously address global warming and the end of the fossil fuel era, are not insignificant, but neither are they unaffordable. However, at the end of 2008, the above is no longer enough:

> The science has moved on. The events the Earth Summit and the Kyoto process were supposed to have prevented are already beginning. Thanks to the wrecking tactics of Bush the elder, Clinton (and Gore) and Bush the younger, steady, sensible programmes of the kind that Obama proposes are now irrelevant. As a Public Interest Research Council report suggests, the years of sabotage and procrastination have left us with only one remaining shot: a crash programme of total energy replacement. [28]

The 'crash programme' could be the ecogrid perhaps – or something like it? If we start soon enough – a big if – the ecogrid project is feasible in purely engineering terms, but it will not remain so for long, as fossil fuels dwindle. Using energy that would otherwise have been frittered away on transport, the prudent and logical course, for mankind, is to secure the limited energy future presented by the ecogrid. With luck, the energy largesse elusively promised by nuclear fusion, deep ocean wave power, and deep sea wind power, can be pursued once the less exotic renewable resources have been harvested by the ecogrid.

> The end does not have to be catastrophic, as long as in the present terminal phase, the last part of the fossil-fuel capital wealth is used to purchase energy conversion machinery for renewable resources. [2]

Research into, and deployment of, an ecogrid will cost considerably more than Tickell's $ 170 billion. Simply on the basis of the man-years of effort, outlined in Sect. 5.3 the project is unlikely to cost less than $ 2.5–3.0 trillion per year. If we add in the remainder of the admirable Kyoto2 disbursement suggestions, a total expenditure in the region of 4% of the global economy is called for. It should be noted that this money does not disappear, like investments in modern banks are prone to do, it gets spent on real physical infrastructure. It represents a good old Keynesian boost to the world economy.

5.4 The Unknowable

From the miniscule scale of nano-engineering, where molecules are massaged to form novel materials, to electrical engineering on the macroscopic scale where massive components are manipulated to build power stations and other large electrical systems, in general the technical problems that are encountered by scientists/engineers, have known or knowable solutions. Because of this most technologists would probably aver that engineering in the physical world is largely ‘child's play’, by comparison with the complexities and imponderables inherent in the social version, where problems seemingly have no definitive answer, or the answers are politically unpalatable, or they have a myriad of answers all of which generate further problems! Unknown knowns or perhaps known unknowns? Who knows? The social engineering of reluctant and recalcitrant mankind, away from its drug-like dependency on fossil fuels, which is absolutely essential to a successful transition to renewable energy, seems a daunting task, if just democratic and market levers are employed. On all the available evidence, it will call for voter and consumer manipulation of a truly Machiavellian quality and scale to shift entrenched attitudes in the market fixated democracies. It is difficult to be confident that this will ever happen. What is really needed is wise leadership. But politicians who are sufficiently knowledgeable to be wise seem no longer to exist.

Returning to the world of do-able engineering, an effective ecogrid, as we have now seen, can be built, but it will be necessary to divert a considerable proportion of the globe's human resources and capital towards implementing this task. The question that then arises is to what extent, and in what way, this huge undertaking will impinge on the rest of the global economy? To answer this kind of broad economic question it is usual to resort to a sophisticated computer model of the economy, but even with the best of these it is prudent to exercise a suitable degree of caution in applying or interpreting the results, because:

> a model is a simplified representation of reality. If it were a perfect replica, it would not be useful. For example, a road map would be of no use to drivers if it contained every feature of the landscape it represents – it focuses on roads and omits, for example, most features of buildings and plants along the way. [29]

At the level of global economics and population dynamics one such model is WORLD3, which is described in detail in ‘Limits to Growth’ (LTG) [29]. This simulation is a complex, non-linear, delayed-response computer program, which keeps track of stocks such as population, industrial capital, persistent pollution, and cultivated land. It calculates the movements of major parameters such as population (through births and deaths), capital stock (through investment and depreciation), arable land (through erosion, pollution, urban and industrial sprawl) and non-renewable resources. All of these parameters and others are linked through multiple nested feedback and feedforward loops. The primary global outputs from the program are presented in three categories, namely: 'state of the world', 'material standard of living', and 'human welfare', as a function of time. The first category includes available resources, food production, industrial output, population level and pollution level. Material standard of living includes life expectancy, food/person, consumer goods/person and services/person, while human welfare presents us with the changes over time in the model 'world' in a human welfare index and in the human ecological footprint.

One of the computer runs described in LTG, directs the WORLD3 model of the global economy to simulate a scenario in which a massive enforcement of pollution reduction is introduced into the model economy. In a 'world', which in 2000 has plentiful non-renewable resources, and in which human beings have developed sophisticated technology to extract these resources, the assumption is made that mankind decides to make a concerted effort to tackle pollution by diverting 4% per year (about $ 2 trillion in 2000) of global output into technology to eradicate it, this effort commencing in 2002. This scenario is not unlike one in which humans might consider diverting 4%/year of global capital into building an ecogrid. The convergence of the two figures is obviously helpful, but it is entirely fortuitous.

The simulation predicts that pollution will continue to rise for nearly 50 years after 2000 despite the drive towards renewables. This is because of delays in implementation and because of continuing industrial and population growth. But, as one would hope and expect, the pollution level is predicted to drop to a considerably lower level from 2030 until 2100, than it would have done in a 'world' where the transition were not attempted (BAU scenario). Pollution never gets high enough to deleteriously affect human health and consequently the transition to renewables succeeds in prolonging the delivery of high welfare to a high population for another generation. But that is it! In the model 'world' the 'good times' come to an end in about 2080, just forty more years further into the future than in the BAU scenario. Population pressure and demand for consumer goods causes a continuing rapid growth in industrial output, soaking up mineral and other resources, so that by 2070 the costs of extraction become so overwhelming to the rest of the economy that collapse occurs. Also, although pollution is lower than in the BAU scenario, it is still enough to have a negative affect on land fertility, and eventually food production cannot match demand from the growing population. Human welfare collapses by about 2080.

The sad fact is that, in the model 'world' of LTG, of the many scenarios, that are postulated and dissected in order to assess possible futures for a resource limited planet, all predict eventual collapse for the global economy. No matter what we do, the end of the 'good times' comes before 2100, *unless* global over-population is seriously addressed. Of course, this is a model prediction, which is no better that the data supplied to it. It should therefore be interpreted with caution. However, as far as one can gauge, virtually nowhere in the political firmament, apart from some laudable but almost undetectable exceptions [30, 31], is there any attempt being made to engage with the issue of population stabilisation, never mind reduction. The Chinese experiment, of limiting families to one child, is generally considered to have been far too draconian! The issue is immensely difficult for the democratic nations of the world because a population policy is a certain vote loser. Its formulation would inevitably entail some rather serious dismantling of irrational and entrenched human belief systems relating to fertility, contraception, abortion, family size, etc., which cannot be done quickly. Unfortunately, time is not on our side, since in the background to all this, is the threat of anthropogenic global warming, which is growing relentlessly. If it is not addressed before scientifically defined tipping points are breached, and a run-away warming process begins, it is not only the global economy that will collapse, but very likely the human species itself.

The unfortunate but unavoidable conclusion is that technology alone, no matter how good, is incapable of providing a cure for a feverish planet. We are seemingly paralysed by an inescapable paradox. The most successful species on the planet is too successful. There are too many of us and we are, as a result, relentlessly despoiling the only habitat we will ever have. Despite the munificence of the planet we inhabit, and despite the opportunities presented to mankind through science and engineering, we appear doomed to fail to grasp them. Mankind's tragedy is that notwithstanding all the possibilities that exist to secure the future, we are unlikely to avail ourselves of them, because we seem trapped by history, beliefs, misconceptions and ignorance – a tragedy, which would have been well understood by that renowned engineering cynic Edward A. Murphy Jr, who coined the tried and tested law (Murphy's law, sometimes referred to as sod's law), which states: 'if anything can go wrong it will', particularly where human beings are implicated [32]. We know this because he is also reputed to have been the source of the aphorism: 'If there are two or more ways (for humans) to do something, and one of those ways can result in a catastrophe, then catastrophe is inevitable'.

It should perhaps be observed that, in fact, in the engineering world at least, Murphy's laws are by no means infallible. They can actually be rendered toothless, by the rigid application, to any given problem, of sound knowledge, good understanding, flawless logic, and unremitting rationality.

Glossary

AC	Alternating current
Ampere (A)	Unit of current
BAU	Business as usual
Bear Stearns	US financial institution
BESS	Battery energy storage system
Billion	One thousand million (1×10^9)
BRIC	Brazil, Russia, India, China
Br	Bromine
Burns (Robert)	Scottish poet (1759–1796)
Bus-bar	Metallic (usually copper) high current interconnector
BTU	British thermal unit (= 1,055 Joules)
°C	Degree centigrade
CAES	Compressed air energy storage
CES	Capacitive energy storage
CH_4	Chemical formula for methane
CNN	Cable News Network (USA)
Coriolis (force)	Inertial force on a moving body caused by the earth's rotation
CO_2	Chemical formula for carbon dioxide
Coulomb (C)	Unit of electrical charge
CSP	Concentrated solar power
DC	Direct current
Dopant	Foreign molecules added to a pure crystal (e.g., silicon to form a semi-conductor)
EC	Electrochemical
ECES	Electrochemical energy storage
Ecogrid	Global electrical power transmission system

EDL	Electric double layer
Electron	Negatively charged sub-atomic particle
emf	Electromotive force
EU	European Union
°F	Degree Farenheit
Farad (F)	Unit of capacitance
Fannie Mae	US federal national mortgage association (FNMA)
FES	Flywheel energy storage
Forcing	Atmospheric warming over and above natural solar warming
Fossil fuels	Carbon based energy sources such as oil, natural gas, and coal.
Freddie Mac	US federal home loan mortgage corporation (FHLMC)
g	Gravitational acceleration for the Earth ($9.81\ m/s^2$)
G8	Abbreviated reference to the 'top' eight global nations
Giga (G)	$\times 10^9$
Great Wall	Ancient defensive wall which criss-crosses China
Greenhouse gas	Mainly carbon dioxide, methane and water vapour
Grid	Electrical power transmission system
HES	Hydrogen energy storage
Henry (H)	Unit of inductance
Hertz (Hz)	Unit of frequency (1 Hz = 1 cycle/second)
HES	Hydrogen energy storage
H_2	Chemical formula for hydrogen
H_2O	Chemical formula for water
HTSC	High temperature superconductor
HVDC	High voltage direct current
Kelvin (Lord)	William Thomson Kelvin (1824–1907) physicist particularly in fields of thermodynamics and electricity
kg	Kilogram
Kilo (k)	$\times 10^3$
Kinetic energy	Energy of motion
LTG	*Limits to Growth*
Mega (M)	$\times 10^6$
MES	Massive energy storage
Micro (μ)	$\times 10^{-6}$
Microwaves	Radio frequencies from 1 GHz to 100 GHz
Milli (m)	$\times 10^{-3}$
m.k.s or MKS	Metre-kilogram-second dimensional system
mm	Millimetre

mph	Miles per hour
Murphy (Edward)	As in Murphy's law
Nano (n)	$\times 10^{-9}$
NBL	Nuclear base load
Newton (N)	Unit of force
NiCad	Nickel–cadmium
Nimbyism	Not in my back yard (-ism)
NT	Non-transport
Ohm (Ω)	Unit of electrical resistance
OWC	Oscillating water column
Pascal (Pa)	Unit of pressure
PEM	Proton exchange membrane
Period	Time occupied by one cycle of a wave
Permian	Geological period 280–250 million years ago
PHES	Pumped hydro-energy storage
Photon	Elementary particle representing the quantum of energy in light – or any electromagnetic wave
Pole	Source (north) or sink (south) for magnetic flux
Potential energy	Energy of position
Proton	Positively charged sub-atomic particle
ppmv	Parts per million by volume
PSD	Passive solar design
PV	Photo-voltaic
Renewables	Sources of energy which are essentially inexhaustible as long as the Sun shines – such as wind.
Robben Island	Prison island off the west coast of South Africa
Scientific American	US Science Magazine
Siemen (S)	Unit of electrical conductance
SMES	Superconducting magnet energy storage
Sputnik	First man-made earth satellite
TEQ	Tradable Energy Quota
Tera (T)	$\times 10^{12}$
Tesla (T)	Unit of magnetic flux density
THES	Thermal energy storage
Tipping point	A climatic or geological event which introduces positive feed back into the global warming process
Third Reich	German regime before and during second world war
Thyristor	Resistive device whose resistance is dependent on the direction of flow of the electrical current
Tonne (metric ton)	= 1000 kg
Triassic	Geological period 250–200 million years ago

Trillion	One million million (1×10^{12})
uhf	Ultra high frequency
UN	United Nations
Valve	Evacuated electrical device which permits current flow through it in one direction only
vlf	Very low frequency
Volt V)	Unit of voltage
Watt (W)	Unit of power
Wavelength	Distance in space occupied by one cycle of a wave
Weber (Wb)	Unit of magnetic flux
Zn	Zinc
ZnBr	Zinc–bromine

References and Notes

Chapter 1

[1] Stott, P.A. *et al.*, Human contribution to the European heat wave of 2003. *Nature* 432: 610–614, Dec. 2004.

[2] Emanuel, K., Increasing destructiveness of tropical cyclones over the past 30 years. *Nature* 436\4:686–688, Aug. 2005.

[3] MacCracken, M.C., Prospects for future climate change and the reasons for early action. *Journal of Air & Waste Management Association,* 50:735–786, 2008.
This article is a clear and comprehensive summary of the global warming issue, which presents and reviews all of the relevant data. A strenuous plea is made for governments to act quickly, with vigour and determination, to engineer an effective strategy for reducing greenhouse gas emissions, but the solutions proffered, ranging from efficiency drives, nuclear build, expansion of solar and wind farms, moves towards a hydrogen economy, carbon capture and replanting of forests, imply that a solution exists within the envelope of global capitalism and the market, and therefore that economic growth and population growth need not be addressed.

[4] Lenton, T.M., Climate change to the end of the millennium. *Climatic Change* 76:7–29, 2006.

[5] http://commentisfree.guardian.co.uk/richard_adams/2006

[6] Ascherson, N., How far can we fall. *The Sunday Herald*, UK, 27 July 2008.

[7] Meadows, D. *et al.*, *The Limits to Growth*. Universe Books, New York, 1972.

[8] Athanasiou, T., *Slow Reckoning*. Secker & Warburg, London, 1996.
In *Slow Reckoning* the source of the ecological threat that mankind is bringing to bear on the planet is attributed to the North/South divide, to the gross differences between the rich and poor world. These differences are being exacerbated by current obsessions with globalisation and markets. The book warns that severe consequences will be experienced if we continue to pursue first world economic strategies to solve global poverty.

[9] Meadows, D. *et al.*, The Limits to Growth: Thirty Year Update. *Earthscan,* UK, 2005.
If you are concerned about the planet and the future of mankind, this is essential reading. By marshalling a huge volume of real and hard data, and feeding it into a powerful software model of the global economy, scenarios for both economic collapse and sustainability are presented.

[10] International Energy Outlook 2008 – Highlights, *Energy Information Administration,* June 2008. http://www.eia.doe.gov/oiaf/ieo/index.html

[11] *Scientific American,* September 2006.

[12] Bartlett, A.A., *The Physics Teacher* 44:623–624, Dec. 2006.

[13] Centre for Alternative Technology, Zero Carbon Britain: An alternative energy strategy. 10 July 2007.

[14] Monbiot, G., *Heat.* Penguin, 2006.

[15] Calhoun, J.B., Death squared: the explosive growth and demise of a mouse population. *Proc. Roy. Soc. Med.* 66:80–88, Jan. 1973.

[16] Brown, L., How food and fuel compete for land. *The Globalist,* Feb. 2006.

[17] Bell, I., The Saturday Essay. *The Herald,* 26 April 2008.

[18] Angel, R. Feasibility of cooling the Earth with a cloud of small spacecraft near the inner Lagrange point. *PNAS* 103(46):17184–17189, 2006.

[19] Budyko, M.I., *Climatic Changes.* American Geophysical Union, Washington DC, 1977.

[20] Latham, J., Control of global warming. *Nature* 347(27):339–340, 1990.

[21] Lovelock, J., *Revenge of Gaia.* Penguin Books, UK, 2006
This book provides an elegant account of the controversial Gaia hypothesis, which views the Earth as a self regulating system, not unlike a living creature. This leads to the conclusion that if the system is pushed too far by an invasive 'disease organism', it may react in forceful and unexpected ways.

[22] Romm, J., *The Hype about Hydrogen.* Island Press, Washington DC, 2004.
Romm takes a hard look at the so called hydrogen economy and finds that it is largely unachievable. This is in a book which is written from a United States perspective. Hydrogen has been widely plugged as a major techno-fix solution, being represented as a primary fossil fuel replacement in the long-term. However, change is considered to be possible only if it is market sympathetic. Consequently, despite dire warnings of the dangers of global warming he sees only very slow advances in the supply of electricity from renewables and envisions a strong road transport sector built around electric vehicles or e-hybrids (Toyota Prius' which can access mains electricity) long after 2030.

[23] Romm, J., *Hell and High Water.* Harper Collins Publishers, New York, 2007.

[24] Tickell, O., *Kyoto 2: How to Manage the Global Greenhouse.* Zen Books, London, 2008.
The Kyoto protocol, and its influence on green house gas emissions in the years since it was ratified, is put under the 'spot-light', with the damning conclusion that it has been little more than 'window dressing'. Carbon trading schemes emanating from Kyoto have made some people a lot of money, but have certainly not had the desired effect of forcing down emissions. More rigorous and hopefully more effective mechanisms, are proposed for a sequel to Kyoto.

[25] Flannery, T., *The Weather Makers*. Penguin Books, 2007.
This is a clear and very readable exposition of the history, the problems and the consequences of climate change. The book is comprehensive encompassing as it does, possible causes of global warming, the growing evidence, and possible solutions, both of the technofix variety and those requiring societal changes. The transport dilemma arising from the lack of an alternative to fossil fuels is addressed, as is the urgent need to procure sustainable electric power.

[26] G8, Breaking the Climate Deadlock. G8 Report, Japan 2008.

[27] Meteorological Office, International Symposium On Stabilisation of Greenhouse Gases: Tables of Impacts, Hadley Centre, Exeter, 2003.

[28] New Economics Foundation, 100 Months. Technical Note, August 2008. http://www.neweconomics.org/gen/uploads/sbfxot55p5k3kd454n14zvyy01082008141045.pdf

[29] Institute of Engineering Technology, Electromagnetics – University and Industry Study. 2002.

[30] RAE, Educating Engineers for the 21st century. The Royal Academy of Engineering Report, 2007 (ISBN 1-903-496-35-7).

[31] Framing the Engineering Outsourcing Debate: Placing the United States on a Level Playing Field with India and China. Duke University, Master of Engineering Program Paper, December 2005.

[32] The Engineering Profession. Engineering Council Report, November 2000.

Chapter 2

[1] Dawkins, R.W., *The Selfish Gene*. Oxford University Press, 1989.

[2] For the mathematically inclined:

> The force on an object situated on, or above, the surface of the earth is given by:
>
> $$F = mg \text{ (N)}$$
>
> where m is mass (kg) and g is the gravitational acceleration (9.81 m/s^2).
> Furthermore, work (potential energy) is force times distance moved in the direction of the force and is
> given by:
>
> $$P.E. = Fd\cos\theta \text{ (J)}$$
>
> where d is the distance moved (m) and θ is the angle (radians) contained by the force vector and the distance vector.
> For a mass moving with velocity v m/s, its kinetic energy is given by:
>
> $$K.E. = \tfrac{1}{2}mv^2 \text{ (J)}$$

[3] For the mathematically inclined:

The power P (W) absorbed or delivered by a time changing system with a instantaneous energy level of W Joules, is given by:

$$P = \pm \frac{dW}{dt} \text{ (W)}.$$

[4] Weinburg, S., *Dreams of a Final Theory*. Vintage, 1993.
A wonderfully lucid explanation of the physics of the day and of the journey toward a 'theory of everything'.

[5] Feynman, R. *et al.*, *Lectures on Physics–II*. Addison Wesley, 1972.
This book contains the most complete explanation of the physics of lightning you are likely to find anywhere.

[6] McKenzie Smith, I., *Hughes Electrical Technology*. Prentice Hall, 1995.
There is a multitude of text books which provide comprehensive instruction on electrical machines. This one keeps it simple.

[7] Chapman, S.J., *Electric Machinery Fundamentals*. McGraw-Hill International, 1985.
This text provides a much more detailed examination of the topic. It is aimed at second/third year undergraduates at a UK university.

[8] Feynman, R. *et al.*, *Lectures of Physics–II*. Addison Wesley, 1972.

[9] For the mathematically inclined:

For an isolated sphere of charge Q (C) in a vacuum, its electric field $\mathbf{E}$ is:

$$\mathbf{E} = \frac{Q}{4\pi\varepsilon_0 R^2}\mathbf{a}_R \text{ (V/m)}$$

where R is radial distance from the charge (m), ε_0 is the vacuum permittivity ($= 8.84\times10^{-12}$ F/m) and $\mathbf{a}_R$ is a unit magnitude radial vector. The force on a charge q placed in this field is given by:

$$\mathbf{F} = q\mathbf{E} \text{ (N)}$$

[10] For the mathematically inclined:

For a long straight conductor carrying a current I (A) the circumferentially directed magnetic field strength at a radial distance R (m) from the axis of the wire is given by:

$$B_\phi = \frac{\mu_0 I}{2\pi R} \text{ (T)}$$

where μ_0 is the vacuum permeability ($= 12.57\times10^{-7}$ H/m).

[11] For the mathematically inclined:

For a stationary wire of length l (m) carrying a current I (A), immersed in a magnetic field of flux density B (T), the Lorentz force is given by:

$$F = B\,I\,l\ (\mathrm{N})$$

F, B and l are mutually orthogonal vectors.

For a straight wire of length l moving with velocity v m/s in a magnetic field B (T):

$$emf = B\,l\,v\ (\mathrm{V})$$

v, B and l are mutually orthogonal vectors. For a rotary generator this becomes:

$$emf \approx 4.15\,T\,\Phi_P\,f\ \ (\mathrm{V})$$

where T = number of turns in the stator winding,

$$\Phi_P = \text{pole flux (webers)} = B_P A_P$$

A_P = pole cross-sectional area (m^2),
f = generated frequency = $N_P \times \mathrm{rpm}/120$ Hz
N_P = number of poles (usually 4).

[12] For the mathematically inclined:

The power lost (W) in a wire of resistance R (Ω) carrying a DC current of I (A) is given by:

$$W = I^2 R\ \ (\mathrm{W})$$

where:

$$R = \frac{\rho\,\ell}{A}\ \Omega$$

Here, ρ is the wire resistivity (Ω.m), ℓ is the wire length (m) and A (m^2) is the wire cross-sectional area.
Note: for an AC current the rms value should be used.

[13] Howell, J.R. and Buckius, R.D., *Fundamental of Engineering Thermodynamics*. McGraw-Hill, New York, 1987.

[14] For the mathematically inclined:

Carnot's Theorem for thermal efficiency (η_{th}) has the form:

$$\eta_{\mathrm{th}} = 1 - \frac{T_{\mathrm{C}}}{T_{\mathrm{H}}}.$$

Consequently for $T_{\mathrm{C}} = 294$ K and $T_{\mathrm{H}} = 1089$ K:

$$\eta_{\mathrm{th}} = \left[1 - \frac{294}{1089}\right] \times 100 = 73\%$$

[15] Angier, N., *The Cannon*. Houghton Mifflin, 2007.
This is simply one of the most elegantly written books on a science topic that I have had the privilege to read.

[16] Krauss, L.M., *Fear of Physics*. Basic Books, New York, 1993.

[17] UK Energy review, http://www.berr.gov.uk/files/file32003.pdf

Chapter 3

[1] Weinburg, S., *Dreams of a Final Theory*. Vintage, 1993.

[2] For the mathematically inclined:

Solar power density $p = 1367\ \text{W/m}^2$
Intercept area $A = \frac{1}{4}\pi D^2 = 1.28\times 10^{14}\ \text{m}^2$
Solar flux $= pA = 1367 \times 1.28\times 10^{14} = 170\times 10^{15}\ \text{W}$

[3] Alfe, D., Gillan, M.J. and Price, G.D., Thermodynamics from first principles: temperature and composition of the Earth's core. *Mineralogical Magazine* 67(1):113–123, 2003.

[4] Steinle-Neumann, G., Stixrude, L. and Cohen, R., *New Understanding of Earth's Inner Core*. Carnegie Institution of Washington, 2001.

[5] World Energy Consumption: www.solcomhouse .com/worldenergy.htm

[6] http://www.ren21.net/globalstatusreport/download/

[7] World Consumption of Primary Energy by Energy Type and Selected Country (XLS). Energy Information Administration, US Department of Energy, 31 July 2006.

[8] Guthrie Brown, J., *Hydroelectric Engineering Practice*, Volume II. Blackie & Son Ltd., 1970.

[9] Aswan High Dam, Web site, http://www.aswandam.net/

[10] Survey of Energy Resources – Hydropower. World Energy Council, 2007.

[11] Gipe, P., *Wind Power*. Chelsea Green Publishing Co., 2004
A plethora of books exist on wind power some going back as far as 1940. This one covers most of the salient material in an easily readable style.

[12] Survey of Energy Resources – Wind. World Energy Council, 2007

[13] For the mathematically inclined:

Kinetic energy is given by: $K.E. = \frac{1}{2}mv^2$ (J)
And power is energy/s Hence:
Wind power $= \frac{1}{2}\rho_A A v^3$ (W)
where ρ_A is the air density (kg/m^3), A (m^2) is the swept area of the turbine blades, and v (m/s) is the wind velocity.

[14] http://www.ocean.udel.edu/windpower/ResourceMap/index-world.html

[15] Count, B., *Power from Sea Waves*. Academic Press, 1980.

[16] Carr, M., Understanding waves. *Sail* 38–45, Oct. 1998.

[17] For the mathematically inclined:

For water waves, wavelength is given by:

$$\lambda = \frac{g}{2\pi f^2} \text{ m}$$

while wave velocity $v = 1.25\sqrt{\lambda}$ m/s.

[18] Survey of Energy Resources – Wave Energy. World Energy Council, 2007.

[19] Kofoed, J.P., Frigaard, P., Friis-Madsen, E. and Sørensen, H.C., *Prototype testing of the wave energy converter wave dragon. Renewable Energy* 31, 2006.

[20] Tedd, J., Kofoed, J.P., Knapp, W., Friis-Madsen, E. and Sørensen, H.C., Wave Dragon, prototype wave power production. *World Renewable Energy Congress – IX*, Florence, Italy, 19–25 August 2006.

[21] Tease, W.K., Lees J. and Hall, A., Advances in Oscillating Water Column Air TurbineDevelopment. EWTEC, 2007.

[22] Wavegen, Final report of OWC development, http://www.wavegen.co.uk/pdf/art.1707.pdf

[23] Wavepower – moving towards commercial viability. ImechE Seminar Publication 2000-8, 2000.

[24] Baker, A.C., *Tidal Power*. Peter Peregrinus Ltd, London, 1991.

[25] G. Boyle (ed.), *Renewable Energy: Power for a Sustainable Future*. Oxford University Press/Open University, 2004.

[26] Survey of Energy Resources – Tidal Energy. World Energy Council, 2007.

[27] For the mathematically inclined:

Area of earth's disc $A_D = \frac{1}{4}\pi D^2$ m^2

where D = earth's diameter.

Surface area of earth $A_S = 4\pi R^2 = \pi D^2$ m^2

[28] Survey of Energy Resources – Solar Energy. *World Energy Council*, 2007.

[29] Cutler, P., *Solid-state Device Theory*. McGraw-Hill Book Co., New York, 1972. A good student-level treatment of semiconductor engineering.

[30] Williams, J.R., *Solar Energy*. Ann Arbor Science Publishers Inc., 1977.

[31] Lorenzo, E. *et al*., *Solar Energy: Engineering of Photovoltaic Systems*. Progensa, 1994.

[32] Blakers, A., Weber, K., Everett, E., Franklin, E. and Deenapanray, S., Sliver cells – a complete photovoltaic solution. IEEE 4th World Conference on Photovoltaic Energy Conversion, Hawaii, May 2006.

[33] Kurokawa, K., Energy from the Desert. http://www.iea-pvps.org/

[34] Arrillaga, J., *High Voltage Direct Current Transmission*. Institution of Electrical Engineers, 1998.

[35] For the mathematically inclined:

Snells laws of reflection and transmission:

1. $$\theta_r = \theta_i$$

where θ_i is the incident angle on a plane interface, and θ_r is the reflection angle.

2. $$\frac{\sin\theta_t}{\sin\theta_i} = \frac{\varepsilon_{r2}}{\varepsilon_{r1}}$$

where θ_t is the transmission angle, while ε_{r1} and ε_{r2} are the relative permittivities in the incident and transmissive media, respectively.

[36] Superheated steam boilers. Spirax Sarco, http://www.spiraxsarco.com

[37] Solar Energy Generating Systems, http://www.flagsol.com/SEGS_tech.htm

[38] Braun, H.W., Solar Stirling gensets for large scale hydrogen production. *Solar Energy Technology* SED-13:21–29, 1992.

[39] Whitford, W.G., Ecology of deserts. *Journal of Mammalogy* 1122–1124, August 2003.

[40] Wohlfahrt, G., Fenstermaker, L.F. and Arnone, J.A., Large annual net CO_2 uptake of a Mojave Desert ecosystem. *Global Change Biology*, http://www.blackwell-synergy.com/doi/abs/10.1111/j.1365-2486.2008.01593.x

[41] Trans-Mediterranean Interconnection for Concentrating Solar Power, *German Aerospace Centre (DLR)*, April 2006. http://www.katharinehamnett.com/images/campaigns/csp_report/TRANS-CSP-REPORT-2006.pdf

[42] Cataldi, R., Hodgson, S.F. and Lund, J.W., Stories from a heated Earth. Geothermal Resources Council and International Geothermal Association, p. 205, 1999.

[43] Geothermal Energy Association, http://www.geoenergy.org/aboutGE/potentialUse.asp#world

[44] Flannery, T., *The Weather Makers*. Penguin, 2007

[45] Survey of Energy Resources – Geothermal Energy. World Energy Council, 2007.

Chapter 4

[1] Electricity Storage Association (www.electricitystorage.org) 2007.

[2] Kurokawa, K., Energy from the Desert. http://www.ieapvps.org/

[3] Herbst, G.E. *et al.*, Huntorf 290MW air storage system energy transfer (ASSET) plant design, construction and commissioning. *Proceedings of the Compressed Air Energy Storage Symposium*, NTIS, 1978.

[4] Crotogino, F., Mohmeyer, K.-U. and Scharf, R. Huntorf CAES: More than 20 Years of Successful Operation. http://www.unisaarland.de/fak7/fze/AKE_Archiv/AKE2003H/AKE2003H_Vortraege/AKE2003H03c_Crotogino_ea_HuntorfCAES_CompressedAirEnergyStorage.pdf

[5] Compressed air energy storage: history. http://www.ridgeenergystorage.com/caes_history.htm

[6] For the scientifically inclined:

The equation of state for all gases under low pressure is:

$$pV = RT$$

where p = pressure (N/m^2), V = volume (m^3), T = Temperature (K), and the universal gas constant $R = 8.314$ J / K for one mole of the gas.

[7] Ter-Gazarian, A., *Energy Storage for Power Systems*. Peter Peregrinus Ltd., 1994.

[8] Katz, D.L. and Lady, E.R., *Compressed Air Storage*. Ulrich's Books Inc., Ann Arbor, Michigan, 1976.

[9] Giramonti, A.J., Preliminary feasibility evaluation of compressed air storage power systems. United Technologies Research Centre Report, R76-952161-5, 1976.

[10] Gill, J.D. and Hobson, M.J., Water compensated CAES cavern design. *Proceedings of the Compressed Air Energy Storage Symposium*, NTIS, 1978.

[11] Renewable energy storage. Imech E Seminar Publication 2000-7, 2000.

[12] Driga, M.D. and Oh, S.-J., Electromagnetically levitated flywheel energy storage system with very low internal impedance. Pulsed Power Conference, 1997. Digest of Technical Papers, 11th IEEE International, 29 Jun–2 Jul 1997 Page(s):1560–1565 vol. 2, issue 1, 1997.

[13] Williams, P.B., Practical application of energy flywheel. *Modern Power Systems* 4(5):59–62, 1984.

[14] Grassie, J.C., *Applied Mechanics for Engineers*. Longmans, Green & Co. Ltd, London, 1960.

[15] Kovach, E.G., Thermal energy storage. Report of the NATO Science Committee Conf., Turnberry, Scotland. Pergammon Press, Oxford, 1976.

[16] Olivier, D. and Andrews, S., *Energy Storage Systems – Past, Present and Future*. Maclean Hunter House, Barnet, UK, 1989.

[17] Wetterman, G. (ed.), *Proceedings of the International Seminar on Thermochemical Energy Storage*, Royal Swedish Academy of Engineering Sciences, Stockholm, 1980.

[18] For the scientifically inclined:

$$NaCl \Rightarrow Na^+ + Cl^-$$

where $\Rightarrow$ implies 'in solution'

[19] Smith, G., *Storage Batteries*. Pitman Publishing Ltd., London, 1980.

[20] For the scientifically inclined:

At the anode:-

$$PbO_2 + 3H^+ + HSO_4^- \Leftrightarrow PbSO_4 + 2H_2O \quad (1.685\text{ V})$$

At the cathode:-

$$Pb + HSO_4^- \Leftrightarrow PbSO_4 + H^+ + 2e^- \quad (0.356\text{ V})$$

For the overall cell:-

$$PbO_2 + Pb + 2H_2SO_4 \Leftrightarrow 2PbSO_4 + 2H_2O \quad (2.041\text{ V})$$

The $\Leftrightarrow$ symbol denotes that the reaction can go both ways. The left to right reaction represents discharge, while charging causes the reaction to go in the opposite direction.

[21] Chang, R., Electrochemistry. In *Chemistry*, 7th edn. McGraw Hill, 2002 (ISBN 0-07-365601-1).

[22] Hart, A.B. and Webb, A.H., Electrical batteries for bulk energy storage. Central Electricity Research Labs., Report RD/L/R1902, 1975.

[23] Talbot, J.R.W., The potential of electrochemical batteries for bulk energy storage in the CEGB system. *Proceedings of International Conference on Energy Storage,* Brighton, UK, 1981.

[24] For the scientifically inclined:

$$xNa + yS \Leftrightarrow Na_xS_y$$

[25] ABB-led Group to Build World's Largest Battery Storage System; Global Power and Technology. *Business Wire*, 29 October 2001.
http://www.allbusiness.com/energy-utilities/utilities-industry-electric-power/6138152-1.html

[26] http://www.earthtimes.org/articles/show/zbb-energy-corporation-unveils-commercial-product-name-for-zinc-energy,287129.shtml

[27] For the scientifically inclined:

At the anode:

$$Zn_{(s)} \Leftrightarrow Zn^{2+}_{(aq)} + 2e^-$$

At the cathode:

$$Br_{2(aq)} + 2e^- \Leftrightarrow 2Br^-_{(aq)}$$

For the overall cell:

$$Zn_{(s)} + Br_{2(aq)} \Leftrightarrow 2Br^-_{(aq)} + Zn^{2+}_{(aq)} \qquad (1.8\text{ V})$$

[28] Winter, C.-J. and Nitsch, J., *Hydrogen as an Energy Carrier: Technology, Systems, Economy*. Springer-Verlag, 1988.

[29] Williams, L.O., *Hydrogen Power*. Pergamon Press, 1980.

[30] Romm, J.J., *The Hype about Hydrogen.* Island Press, Washington, 2005.

[31] For the scientifically inclined:

First reaction:
$$CH_4 + H_2O \Rightarrow CO + 3H_2$$

Second reaction:
$$CO + H_2O \Rightarrow CO_2 + H_2$$

[32] Dooley, J. and Wise, M., Why injecting CO_2 into various geological formations is not the same thing as climate change mitigation: the issue of leakage. College Park MD: Joint Global Climate Change Research Institute [Battelle-Pacific Northwest Nat'l Lab.], 2002.

[33] Hawkins, D., Passing gas: policy implications of leakage from geological carbon storage sites. National Resources Defence Council, Washington DC, 2002.

[34] For the scientifically inclined:

At the anode of the electrolysis cell:
$$2OH^- \Rightarrow O_2 + H_2O + 2e^-$$

At the cathode:
$$2e^- + 2K^+ + 2H_2O \Rightarrow 2K^+ + H_2 + 2OH^-$$

[35] Bockris, J. O'M., *Energy Options.* Taylor & Francis Ltd., 1980.

[36] Sperling, D. and Cannon, J.S., *The Hydrogen Energy Transition.* Elsevier Academic Press, 2004.

[37] McKenzie Smith, I., *Hughes Electrical Technology.* Prentice Hall, 1995.

[38] For the mathematically inclined:

For a capacitor of capacitance C (F), charge stored is given by:
$$Q = CV \text{ (C)}$$
where V is the voltage between the plates (V). Furthermore:
$$C = \frac{\varepsilon A}{d} \text{ (F)}$$
where $\varepsilon = \varepsilon_r \varepsilon_0$ is the absolute permittivity (F/m) of the material separating the plates, while ε_r is the relative permittivity; A is the area of the plates (m^2) and d is the plate separation (m).

[39] For the mathematically inclined:

The energy stored in a capacitor (W_e J) is given by:
$$W_e = \tfrac{1}{2}CV^2 \text{ (J)}$$

[40] Angier, N., *The Cannon.* Houghton Mifflin, 2007.

[41] Belyakov, A.I., Capacitor with double electric layer. US Patent No. 5,923,525, 13 July 1999.

[42] Yoshio, M. and Nakamura, H., Power storage element and electrical double-layer capacitor. European Patent No. EP1727166, 29 November 2006.

[43] Bendler, J.T. and Takekoshi, T., Molecular modeling of polymers for high energy storage capacitor applications. *IEEE 35th International Power Sources Symposium*, 22–25 June 1992, pp. 373–376.

[44] For the mathematically inclined:

For an infinitely long straight conducting wire carring current I (A), Ampere's law gives:

$$H = \frac{I}{2\pi R} \quad \text{(A/m)}$$

where R (m) is the radius from the axis of the wire.

For a coil of large length to diameter ratio, Ampere's law gives:

$$H = \frac{NI}{\ell} \quad \text{(A/m)}$$

where N = number of turns, I = current (A), and ℓ = coil length (m)

[45 For the mathematically inclined:

The energy stored in a coil carrying a current I (A) is:

$$W_h = \frac{1}{2} L I^2 \quad \text{(J)}$$

where L is the inductance (H). Also

$$L = \frac{\ell}{\mu A} \quad \text{(H)}$$

where ℓ is the coil length (m), A is the cross-sectional area (m^2), and $\mu = \mu_0 \mu_r$ is the magnetic permeability of the core of the coil.

[46] Tinkham, M., *Introduction to Superconductivity*, 2nd edn. Dover Books on Physics, 2004. ISBN 0-486-43503-2

[47] Maddock, B.J. and James, G.B., Protection and stabilisation of large superconducting coils. *Proceedings of IEE* 115(4):543–547, 1968.

[48] Wisconsin superconductive energy storage project. University of Wisconsin, Madison, USA, 1974–1976.

[49] Andrianov, V.V. *et al.*, n experimental 100MJ SMES facility. *Cryogenics* 30:794–798, 1990.

[50] Lovelock, J., *The Vanishing Face of Gaia*. Basic Books, 2009.
This follow up book to *The Revenge of Gaia* is a disturbing read, insofar as the author makes it clear that it is probably too late for mankind to arrest global warming. Attempting to reduce greenhouse gases by efficiency measures and by the adoption of renewables is deemed to be futile. Instead the book recommends that governments should be concentrating on the adaption of their real estate to accommodate rising seas, a hostile climate, and massive population migration towards the poles.

[51] Hewitt, G.F. and Collier, J.G., *Introduction to Nuclear Power*. Taylor & Francis, 2000

[52] www.wise-uranium.org/stk.html

[53] OECD, Uranium 2005: Resources, Production Demand. OECD: International Atomic Energy Agency (IAEA), 2 June 2006.

[54] Trans-Mediterranean Interconnection for Concentrating Solar Power, German Aerospace Centre (DLR), April 2006.

Chapter 5

[1] World Energy Consumption: www.solcomhouse.com/worldenergy.htm

[2] Brockris, J.O'M., *Energy Options*. Taylor Francis Ltd., London 1980.

[3] Tickell, O., *Kyoto 2: How to Manage the Global Greenhouse*. Zen Books, London, 2008.

[4] Breaking the Climate Deadlock. G8 Report, Japan 2008.

[5] The Future of Nuclear Power. Massachusetts Institute of Technology, 2003. web.mit.edu/nuclearpower/

[6] www.wise-uranium.org/stk.html

[7] http://en.wikipedia.org/wiki/Human_powered_vehicle

[8] http://www.sustainableliving.com.au/profiles/stories/solar-powered-lawn-mower/

[9] Bressand, F. *et al.*, Curbing global energy demand growth:the energy productivity opportunity. McKinsey Global Institute, May 2007.

[10] Romm, J.J., *The Hype about Hydrogen*. Island Press, Washington, 2005.

[11] Horton, G., Forecasts of CO_2 emissions from civil aircraft for IPCC. Qinetic Report No. 06/2178, 2006. (www.berr.gov.uk/files/file35675.pdf)

[12] Synthesis Report of the IPCC, 4th Assessment Report, IPCC, Nov. 2007.

[13] Monbiot, G., *Heat*. Penguin, 2006

[14] Hillman, M., *How We Can Save the Planet*. Penguin, 2004. ISBN 0-14-101692-2

[15] http://gow.epsrc.ac.uk/ViewGrant.aspx?GrantRef=EP/G001634/1

[16] Nissan to launch solar powered city car. *Which?*, September 2008. http://www.which.co.uk/news/2008/09/nissan-to-launch-solar-powered-city-car-156075.jsp

[17] Airship of the Future. Aeroscraft Corp., 2008 (www.aeroscraft.com)

[18] Dossas, V. and Kraft, C., Hydrogen lighter-than-airship. US Patent 6896222, 24/5/2005.

[19] Veldhius, I. *et al.*, A hydrogen fuelled gas-turbine powered high-speed container ship. *International Conference on Fast Sea Transport*, St. Petersburg, Russia, 2005.

[20] Sun, wind, fuel cells, power cargo ship of the future. www.peopleandplanet.net, April 2005.

[21] McWhirter, I., Rising from the ashes. *Sunday Herald*, Scotland, 19 October 2008.

[22] www.autoindustry.co.uk/statistics/

[23] Braun, H.W., Solar Sterling gensets for large-scale hydrogen production. *Solar Energy Technology* SED-13:21–29, 1992.

[24] Kurokawa, K., Energy from the Desert. http://www.iea-pvps.org/

[25] James, W., *The Moral Equivalent of War. Essays of Faith and Morals*. Longman Green, New York & London, 1949.

[26] Athanasiou, T., *Slow Reckoning*. Secker & Warburg, London, 1996.

[27] Hansen, J. *et al.*, Target atmospheric CO_2: where should humanity aim. *eprint arXiv:0804.1126,* April 2008.

[28] Monbiot, G., *The Guardian Newspaper*, UK, 25 November 2008.

[29] Meadows, D. *et al.*, *The Limits to Growth: Thirty Year Update*. Earthscan, UK, 2005.

[30] Lovelock, J., *The Vanishing Face of Gaia*. Basic Books, 2009.

[31] Porritt, J., November 14, 2008, http://www.jonathonporritt.com/pages/2008/11/population_1.html

[32] http://www.xs4all.nl/~jcdverha/scijokes/9_6.html

Index

E

F

O

P

Q

R

Y

Z